D1676289

A Probus Guide to World Markets

QUANTITATIVE INTERNATIONAL INVESTING

A Handbook
of Analytical and
Modeling Techniques
and Strategies

BRIAN R. BRUCE

McGRAW-HILL BOOK COMPANY

London • New York • St. Louis • San Francisco • Auckland
Bogotá • Guatemala • Hamburg • Lisbon • Madrid • Mexico
Montreal • New Delhi • Panama • Paris • San Juan • São Paulo
Singapore • Sydney • Tokyo • Toronto

Published in the United Kingdom by
McGraw-Hill Book Company (UK) Limited
Shoppenhangers Road
Maidenhead, Berkshire England SL6 2QL

UK and European sales and distribution rights to this book are owned by McGraw-Hill Book Company (UK) Limited

British Library Cataloguing in Publication Data

Bruce, Brian R.
Quantitative international investing: a handbook of analytical and modeling techniques and strategies.
1. Foreign investment
I. Title
332.673

ISBN 0-07-707366-5

This publication is designed to provide accurate and authoritative information in regard to the subject matter covered. It is sold with the understanding that the publisher is not engaged in rendering legal, accounting or other professional service.

Printed in the United States of America

1 2 3 4 5 6 7 8 9 0

Dedication

To Susan and Robbie

Additional Titles in
The Probus Guide to World Markets Series . . .

Innovation and Technology in the Markets: A Reordering of the World's Capital Market Systems, Daniel R. Siegel, Editor

Forthcoming Titles

The European Options and Futures Markets: An Overview and Analysis for Money Managers and Traders, The European Bond Commission

The Global Bond Markets: State-of-the-Art Research, Analysis and Investment Strategies, Jess Lederman and Keith K. H. Park, Editors

The Global Equity Markets: State-of-the-Art Research, Analysis and Investment Strategies, Jess Lederman and Keith K. H. Park, Editors

Managing Foreign Exchange Risk: Strategies for Global Portfolios, David F. DeRosa

Contents

About the Contributors

Robert A. Brown received an A.B. in mathematics and economics from Oberlin College, a M.A. in economics from the University of Maryland, College Park, and a M.A. in economics and a Ph.D. in finance from Northwestern University. Dr. Brown also holds the CFA designation.

Prior to joining Ibbotson Associates, Dr. Brown worked as a Commercial Lines Insurance Underwriter for the St. Paul Companies, a Professor of Investments at the University of Colorado, a consultant for IBM Corporation and most recently as the Vice President of the portfolio management group for United Banks of Colorado. He has experience in the design, implementation and marketing of domestic fixed-income and both domestic and foreign equity investment products.

Dr. Brown takes a lead role in consulting on asset allocation projects and general quantitative product development for Ibbotson Associates. He conducts advanced asset allocation seminars covering simulation and multi-period planning techniques.

Brian R. Bruce is a Vice President in the Asset Management Division of State Street Bank and Trust Company. His responsibilities include managing passive international portfolios.

Mr. Bruce is also Editor of *Investing* magazine. He frequently contributes articles to professional journals and speaks at investment conferences.

Previously, Mr. Bruce was Manager of International Equity Products at Northern Investment Management Co., a subsidiary of The Northern Trust Company. He managed global equity portfolios, created a tactical asset allocation model with Roger Ibbotson and developed computer models for portfolio management. Before joining Northern Trust, he was with Stein Roe & Farnham and Eastman Kodak.

Mr. Bruce received an M.S. in computer science from De Paul University and his M.B.A. from the University of Chicago, where he won the CEO Award.

Thomas Clark is a Principal in the International Index and Portfolio Trading Department at Morgan Stanley where he has worked for ten years. He has an M.B.A. from Harvard University.

H. Gifford Fong is President of Gifford Fong Associates, a California investment technology consulting firm with a specialty in fixed income portfolio analysis and asset allocation strategies.

He is a graduate of the University of California, where he earned his B.S., M.B.A. and law degrees.

Mr. Fong is on the editorial boards of *The Journal of Portfolio Management* and *The Financial Analysts Journal* and is a contributor to the Chartered Financial Analysts' sponsored text on portfolio management, "Managing Investment Portfolios: A Dynamic Process." He is a member of the Board of Directors and Program Chairman of the Institute for Quantitative Research in Finance; and a member of the editorial advisory board for *The Handbook of Fixed Income Securities*. He was formerly Institutional Director of the Financial Management Association, and a former member of the advisory board for the Investment Technology Association, New York. He is a contributor to a number of professional books and journals.

In addition, Mr. Fong is co-author of *Fixed Income Portfolio Management*. a book published by Dow Jones-Irwin.

As of June 1986, Gifford Fong Associates became a subsidiary of the Prudential Insurance Company of America.

Michael Gibson is an associate at Morgan Stanley. He has an M.B.A from UCLA.

Steven L. Fradkin is a Second Vice President in the International Services Division of the Corporate Financial Services Group at The Northern Trust Company, Chicago.

Prior to joining the International Services Division in 1987, Mr. Fradkin served as a Senior Analyst and Trust Officer in the Legal, Legislative and Regulatory Affairs Group servicing Master Trust clients.

Mr. Fradkin joined The Northern Trust in 1985 after graduating with a B.A. in economics from Washington University in St. Louis. He is currently enrolled in Northwestern University's J.L. Kellogg Graduate School of Management. He is a member of the Association of Canadian Pension Management, the Canadian Pension Conference and the Chicago Council of Foreign Relations.

Mr. Fradkin will be relocating to The Northern Trust's branch office in London, England early in 1990 to assume responsibility for sales and marketing of The Northern Trust's Master Trust and Global Custody services to U.K. and European clients.

Philip Green manages foreign exchange portfolios for the Investment Management Group of Bankers Trust Company. In addition, he specializes in developing and managing other quantitative investment strategies including liquidity premium capture strategies and option strategies.

Phil has a B.S.E. in International Business and Finance from The Wharton School of Business.

Roger G. Ibbotson is a professor of finance at the Yale School of Management. Professor Ibbotson earned his B.A. at Purdue University in 1965, his M.B.A. at Indiana University in 1967 and his Ph.D. from the University of Chicago in 1973. His book with Rex A. Sinquefield, *Stocks, Bonds, Bills, and Inflation,* is the authoritative source of historical capital market returns in the United States. At the University of Chicago, he managed the university's bond portfolios while completing his dissertation on the unusual price performance of common stock new issues.

Professor Ibbotson is a businessman as well as an academic. His consulting firm, Ibbotson Associates, specializes in applying the tools of academic financial theory to portfolio management, corporate finance and legal and regulatory issues. The Products Division of Ibbotson Associates disseminates returns and other data on the stock, bond, cash, real estate, foreign exchange and other capital markets of the world. Ibbotson Associates was founded in 1977 and currently employs twenty-five people in its Chicago and New Haven offices.

William W. Jahnke is Chairman of Vestek Systems, a leading provider of investment management information systems. Prior to cofounding Vestek in 1983, Mr. Jahnke was Senior Vice President and Manager, Wells Fargo Investment Advisors, in charge of trust and investments. He joined Wells Fargo in 1969 after receiving his M.B.A. from the University of California, Berkeley. In 1967, Mr. Jahnke received an undergraduate degree in Economics from Stanford University.

Mr. Jahnke has spent most of his career developing and applying quantitative investment management tools. In 1973 and 1974, he received a Graham and Dodd Award for articles published in *The Financial Analysts Journal.*

Seth M. Lynn before founding Great American Core Investors in 1984, Seth Lynn was Vice President and head of the Special Products and Technology Division of the Investment Management Group of Bankers Trust in New York. He was responsible for over $13 billion in passively managed assets. Mr. Lynn received a B.A. degree from Yale College and an M.B.A. from the Wharton School, University of Pennsylvania.

John Meier received his B.S. in Chemical Engineering from Michigan State University. Following graduation, he spent eight years in various engineering positions with Standard Oil Company of Ohio (SOHIO). Most of his time was spent directing engineering studies and initial design efforts on expansions to SOHIO's Prudhoe Bay, Alaska, oil field.

He returned to school and received his M.B.A. with an emphasis in finance from the University of California, Berkeley. Mr. Meier joined BARRA in August 1988 as a consultant, International Services. He has been involved in product development, client support and research on BARRA's non-U.S. equity, global equity, global bond and asset allocation risk models. He has been particularly active in BARRA efforts in the fields of international portfolio performance attribution, multiple international manager analysis and international normal portfolios.

Laurence B. Siegel managing director of the Chicago-based investment and economic research firm Ibbotson Associates, earned his B.A. (1975) and M.B.A. in finance and economics (1977) from the University of Chicago. He is a widely known author and speaker on capital market history and forecasts, the cost of capital, asset allocation, investment strategy and policy and other topics relevant to asset management. He serves on the advisory board of *The Journal of Portfolio Management* and other organizations.

Mr. Siegel's consulting practice, developed through more than a decade of affiliation with Ibbostson Associates, applies these fields of knowledge to investment management and corporate finance. Among his principal clients in the investment field are asset management firms, investment banking firms and pension plan sponsors. He has also testified extensively in writing and orally in legal and regulatory ratemaking settings.

Lee R. Thomas III is a Member of the Management Committee of Investcorp Bank E.C. Investcorp, and international investment bank that operates out of offices in London, New York and Bahrain, invests in corporations, real estate and liquid securities, and offers them to individuals, corporations and institutions worldwide, Dr. Thomas is responsible for the firm's proprietary trading, and also manages Investcorp's open-ended Foreign Exchange Fund.

Prior to joining Investcorp, Dr. Thomas was an Executive Director of Goldman Sachs International, Ltd. His work on international investing has been published in many books and journals, including the *Financial Analysts Journal, The Journal of Portfolio Management, Investment Management Review*, the *Journal of Futures Markets*, the *International Journal of Forecasting*, and the *Journal of International Securities Markets*. He received his Ph.D. from Tulane University.

Robert G. Tompkins is Managing Director of the Minerva Group, an international training and consulting firm with offices in the United Kingdom, West Germany, Italy, Spain, Finland and Australia. In two years, over forty clients have retained Minerva to train staff on financial markets and establish dealing operations. Before forming Minerva in late 1987, Mr. Tompkins was the Futures and Options Specialist for Merrill Lynch, London. In that capacity, he was responsible for training clients about these product areas and providing a number of seminars to major European financial institutions, including Central Banks. Prior to joining Merrill, Mr. Tompkins was located in Chicago; initially at the Chicago Mercantile Exchange Research Department where he was involved in the development of financial futures and options markets. Later, he traded derivative products for the portfolio of Harris Bank, Chicago, and was responsible for risk management of Interest Rate Products at Continental Illinois. In addition to his practical experience, he has a B.A., M.A. and M.B.A. (with honors) from the University of Chicago. He is presently at the London School of Economics where he is pursuing a Ph.D. in Finance and Accounting. Mr. Tompkins has published ten articles on topics related to financial markets and has presented more than 140 lectures in 28 countries.

Heydon Traub is a Vice President in the Global Enhanced Equity Department at State Street Bank and Trust Company. His responsibilities include the development of quantitative stock selection and country selection models and currency hedging strategies. He holds an undergraduate degree from Brandeis University and an M.B.A. from the University of Chicago. He is a Chartered Financial Analyst.

Peter Vann is Manger, Risk Management, at Westpac Investment Management Pty Limited. He recently joined Westpac following over two years with a major investment bank. In the last twelve months of that period he worked on security price behavior and characteristics and published twenty-one papers and speeches ranging from futures and derivative strategies to asset allocation and trading strategies, primarily in the areas of fixed income and equities.

He has contributed significantly to the development of market indexes in debt instruments, including liability profiles. He recently acted in a consulting manner to a major Australian State Government Treasury to design a debt liability management and benchmark structure.

At Westpac, his major activity is to bring Investment Management to "state of the art" in risk management and technology by world standards. This involves product vision and creation, development, implementation and selling to institutional investors.

Mr. Vann holds a B.Sc., a M.Sc.St. and a Ph.D.

Marvin B. Waring manages the consulting efforts of Ibbotson Associates, Inc., a Chicago-based investment and economic research and consulting firm. His work is concentrated on new investment product development, investment strategy research, cost of capital, project evaluation and management consulting for the money management and corporate finance communities. He also testifies in securities and commercial litigation on issues surrounding the calculation of damages where financial issues are involved.

Mr. Waring earned his B.S. in economics from the University of Oregon (1971), his J.D. from Lewis and Clark University (1975) and his M.P.P.M. with a concentration in finance and management from Yale University (1987).

CHAPTER 1

Introduction

Brian R. Bruce
State Street Bank and Trust Company

Why should you read this book? We hope this introduction will provide some insight into the thought process that went into assembling this collection.

There are many articles and books on the subject of money management. There are very few books on the basics of how to manage funds. Looking, in any bookstore you can find "how to" books on home repairs, cooking, and self help. If we move past "Thin Thighs in 30 Seconds," "How to Fricassee Hamburger Helper," and "Building Your Dream House on Your Lunch Hour," we find "how to" books on investing. Virtually all of them concentrate on personal finance ("Make Trillions in Real Estate with No Money Down" and "Donald trump Guide to Sound Investing"). Our goal with this book is to teach the professional investor the basic foundation needed to manage international quantitative funds.

We have brought together a group of authors from some of the largest money management firms in the world to create this book. You should come away with an understanding of what issues face fund managers every day.

The book covers both equity and fixed income along with other topics relevant to anyone that manages funds. Within each asset class (international equity and international fixed income), there are several generic styles of management ranging from those using large amounts of judgment to those using little active judgment. The judgmental or active funds are typified by security selection. Less judgmental funds are a combination of active and passive strategies known as active/passive or quantitative funds. These funds combine active judgments with the discipline of passive techniques. It is in this area where the book begins it's coverage. We will also cover the core passive strategies and ways to enhance them.

Our first section begins with a study of the international equity and fixed income asset classes. For anyone who is already familiar with the reasons for diversifying your portfolio into the international arena, these chapters can be skimmed over without losing any content. We move into the passive management of equity portfolios with a chapter from Heydon Traub of State Street Bank, the leading international index firm. I had hoped to follow with a chapter about passive international bond management, however, it is still in it's infancy as the book is being put together and no one is willing to share their secrets yet. Hopefully, we will have a chapter in a future volume. Next, Seth Lynn covers a quantitative strategy using ADR's.

The section moves on to cover both equity and fixed income portfolios enhanced through the use of derivatives before Lee Thomas talks about hedging both asset classes. Lee is one of the leading experts in the world on this subject. Before we leave the section we cover what is the nuts and bolts on investing, how to trade a passive portfolio. It is suprisingly different from trading an individual or active manager would do. We are lucky to have one of the leading global program trading firms, Morgan Stanley, contributing this effort.

The next section consists of five chapters that should be of interest to anyone who manages an international portfolio. John Meier from BARRA has written an excellent chapter on how to find an appropriate benchmark from which to judge your performance. The Northern Trust, voted the number one global custodian in 1988 and 1989, writes on what you can expect from your custodian and services to be aware of. The section follows with a chapter from Bill Jahnke taking us back to the origins of quantitative management, this is must reading for any student of the field, followed by a chapter from Bankers Trust that looks at ways to potentially enhance your portfolio through foreign exchange management and ends with Gifford Fong looking at asset allocation. You will have spent most of this book looking down into two asset classes in detail and we end by looking back up to how the asset classes interact.

Our goal has been to make this the best work possible. We hope you find it more useful than "101 Uses for a Dead Cat," "The Memoirs of Pee Wee Herman" or "1000 Readings in U.S. Tax Law." We wish you the very best in your career in investments!

SECTION I

HOW TO MANAGE EQUITY AND FIXED INCOME

CHAPTER 2

Introduction to International Equities

Roger G. Ibbotson
Ibbotson Associates, Inc.

Laurence B. Siegel
Ibbotson Associates, Inc.

Marvin B. Waring
Ibbotson Associates, Inc.

The Setting

The world equity market has grown phenomenally in the past generation to a size approximating $7 trillion. In 1960 the stock markets of one country—the United States—dominated the world, representing two-thirds of world equity market value. Today the Japanese market is larger than that of the United States, and European and other markets have also grown greatly in size and importance. Figure 1 compares the growth of equity capitalization in the United States and foreign markets.

Part of the growth of the foreign markets has come from new issues. More important, rising share prices have driven market capitalizations upward. Foreign markets have, on average, outperformed that of the United States in the last 30

* The authors wish to thank Paul Kaplan. While we realize it is poor form to thank one's editor, Brian Bruce acted in the capacity of coresearcher for this chapter, and we are grateful to him for that effort.

Figure 1 **Market Capitalizations of U.S. and Foreign Equity Markets, 1960–88 (values in billions of U.S. dollars)**

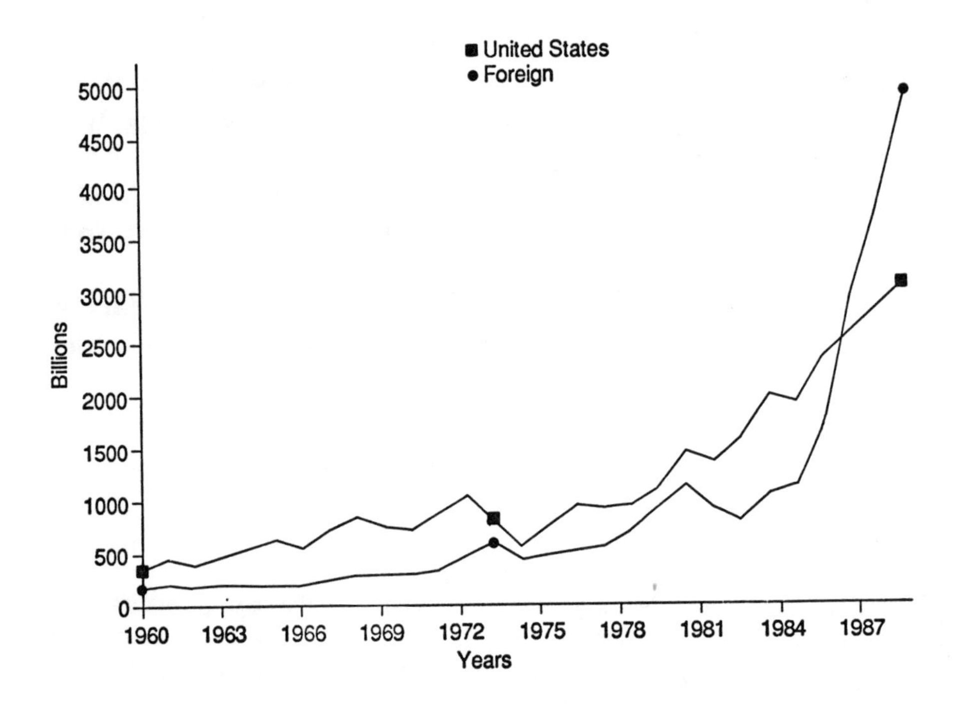

years, creating a vast area of opportunity for the U.S.-based investor not only to diversify but to earn high returns.

As the world markets have grown, they have also become more integrated. Cross border transactions (wherein an investor based in one country transacts in shares of a company based in another country) have become commonplace. Many foreign firms are listed on the New York Stock Exchange, and a great many more foreign (non-British) firms are listed on the London exchange. Such cross border activity is quite new. Traditionally, only continental European investors scaled the barriers of the international markets; they did this out of need, in order to diversify out of their small domestic markets. (Even British investors tended to concentrate on British stocks until recently.) Today the barriers are coming down, and investors are recognizing that even a large country like the United States is somewhat undiversified in the context of a world economy. The future of international investing thus looks even brighter than the present.

Stock Exchanges Around the World

As in the United States, most of the market value of foreign securities trading takes place on organized exchanges, not over the counter. We can thus characterize the markets of the world, in a more or less satisfactory fashion, by looking at the attributes of the stock exchanges. Table 1 presents data for exchanges in 32 countries. Note that while the large, developed countries account for the great majority of market capitalization and trading, stock market activity is found in almost every non-Communist country. There are additional, small markets not shown in Table 1.

Stock exchanges—in the sense of centralized auction markets—do not tell the whole story. In the United States, the National Association of Securities Dealers has set up a National Market System (the NASDAQ NMS), which gives some of the qualities of an exchange to the over-the-counter market. Advances in technology cause this king of decentralized, electronic exchange to have certain advantages over traditional exchange floors. The NASDAQ NMS has been gaining in relative market share, and parallel efforts in other countries can be expected to thrive, too.

Size of the Markets

The sizes of the stock markets of the world are depicted in Figure 2. Size is defined as the sum of the market capitalizations (prices times number of shares outstanding) of the stocks in a market. Japan and the United States are by far the largest markets, each with over $3 trillion in market capitalization. The size of the Japanese market is overstated because of crossholdings (one company holding the stock of another). After correcting for crossholdings, however, that market is still unusually large relative to Japan's population and GNP.

The European market is also large, totaling about $1 1/2 trillion, but is fragmented because of the large number of countries it comprises. Of the European countries, Britain has the lion's share of equity value, although the economies of Germany and France are larger as measured by GNP. The prominence of the British stock market has been attributed to an equity culture, common to the English-speaking countries and Japan, in which individuals and their agents (chiefly pension funds) have a capitalist bent and a correspondingly strong demand for corporate equities.

Table 1 Stock Exchanges Around the World

Country	Principal Exchange(s)	12/31/87 Market Capitalization ($ billions)	1987 Trading Volume ($ billions)	1987 Number of Issues Listed		Principal Market Indexes
				Domestic	Total	
Australia	Sydney plus 5 regionals	137.4	60.4	n/a	1197	All Ordinaries—324 issues
Austria	Vienna	7.5	1.9	96	138	GZ Aktienindex—25 issues
Belgium	Brussels	41.3	11.7	191	331	Brussels Stock Exchange index—192 issues
Brazil	Rio de Janeiro	15.6	n/a	n/a	658	IBV (Rio de Janeiro Stock Index)
Canada	Toronto[1]	76.8	n/a	1628	1695	TSE 300 Composite Index
Chile	Santiago	4.9	0.49	n/a	226	
Denmark	Copenhagen	20.1	2.4	n/a	281	Copenhagen Stock Exchange Index—38 issues
Finland	Helsinki	20.8	6.2	102	105	
France	Paris plus 6 regionals	153.6	182.6	482	677	CAC (Compagnie des Agents de Change)—249 issues
Germany	Frankfurt[2]	190.0[3]	283.3	281	522	DAX; FAZ (Frankfurter Allgemeine Zeitung—100 issues
Greece	Athens	1.2	.04	n/a	114	
Hong Kong	Hong Kong	54.0	47.8	273	288	Hang Seng Index—33 issues
India	New Delhi[4]	8.4	.14	n/a	2180	
Ireland	London, England[5]					
Italy	Milan[6]	161.7	56.6	n/a	286	Banca Commerciale—209 issues
Japan	Tokyo[7]	2724.2	2028.6	1533	1621	TOPIX—1097 issues; TSE-II —423 issues; Nikkei 225
Korea	Seoul	15.0	12.0	n/a	355	Korea Composite Stock Price Index
Malaysia	Kuala Lumpur	25.9	1.4	n/a	301	Kuala Lumpur Stock Exchange Industrial Index
Mexico		2.3	1.6	n/a	191	Indicatores Mexicana—48 issues

(Table continues)

Country	Principal Exchange(s)	12/31/87 Market Capitalization ($ billions)	1987 Trading Volume ($ billions)	1987 Number of Issues Listed		Principal Market Indexes
				Domestic	Total	
Netherlands	Amsterdam	86.1	85.7	283	573	ANP-CBS General Index—51 issues
New Zealand	Wellington	13.5	1.4	318	416	
Norway	Oslo	11.9	8.8	149	156	Oslo Bors Stock Index—50 issues
Portugal	Lisbon[8]	.82	.05	n/a	40	Bank of Portugal index
Singapore	Singapore	42.7	4.9	127	321	Straits Times index—30 issues; SES—32 issues
South Africa	Johannesburg	89.4	6.3	n/a	n/a	JSE Actuaries Index
Spain	Madrid[9]	58.4	7.6	n/a	312	Madrid Stock Exchange Index—72 issues
Sweden	Stockholm	48.9	21.3	164	171	Jacobson & Ponsbach—30 issues; Affarvarkdem General—30 issues
Switzerland	Zurich[10]	155.2	41.2	254	454	Societe de Banque Suisse
Taiwan	Taipei	18.6	23.0	n/a	145	
Thailand	Bangkok	2.0	.61	n/a	97	
U.K.	London	674.1[11]	386.5	1911	2577	Financial Times (FT) Ordinaries—750 issues; FTSE 100; FT 30
United States	New York[12]	2216	1873.0	2174	2244	S&P 500; NYSE Composite; Dow Jones Industrial Average

1 Regional exchanges at Montreal and Vancouver.
2 Six regional exchanges also play a significant role.
3 Domestic issues only.
4 Numerous regional exchanges play a role in the market.
5 Irish shares are traded on the International (London) Stock Exchange, and data for the Irish markets are commingled with United Kingdom data.
6 Numerous regional exchanges play a role in the market.
7 Osaka and other regional exchanges play a significant role.
8 Official and unofficial markets combined.
9 Regional exchanges at Barcelona and Bilbao have significant volume.
10 Exchanges at Basle and Geneva list most of the same stocks as the Zurich exchange and have significant volume. Regional exchanges are at Bern and St. Gallen.
11 Domestic shares only. Total market capitalization, including foreign shares, is $1,997.1 billion.
12 The American Stock Exchange, NASDAQ, and regional stock exchanges are very important. For brevity, data on these trading venues are excluded here.

Figure 2 World Equities: Market Capitalizations (total value approximately $10 trillion)

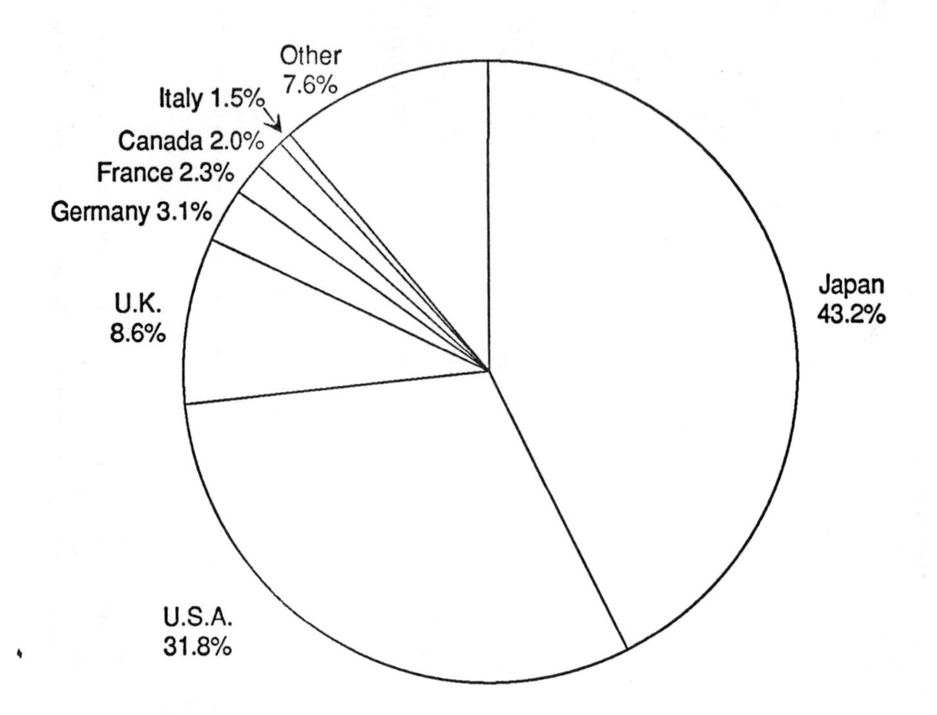

Some of the most interesting and high-returning markets are small and do not appear in Figure 2. These are the emerging markets, situated in countries that are attempting the transition from Third World to First World. Korean and Taiwan are the largest markets in the emerging group, each with a market capitalization a little smaller than that of Denmark. Special barriers exist, however, in some emerging markets; for example, it is difficult for non-Koreans to purchase Korean stocks. Investors often capture returns in emerging markets by holding shares of single-country mutual funds (country funds) or of mutual funds that diversify across emerging markets.

Historical Returns

Historical returns in the various stock markets range from spectacular to dismal. Of the countries with 20 years of data as shown in Table 2, Japan had the highest return in local currency terms, with a compound annual rate of 18.3 percent. The Swiss market fared the worst, with a compound annual return of 4.2 per-

Table 2 Summary Statistics of Annual Returns on World Equities: Total Returns in Local Currencies

Country	Period	Compound Annual Return (%)	Arithmetic Mean Return (%)	Standard Deviation of Returns (%)
Australia	1969–88	10.46	13.98	29.37
Austria	1969–88	7.08	9.06	24.47
Belgium	1969–88	15.73	17.32	19.56
Canada	1969–88	10.89	12.17	17.13
Denmark	1969–88	15.03	19.91	38.22
France	1969–88	14.16	21.70	44.11
Germany	1969–88	7.93	10.56	25.05
Hong Kong	1970–88	18.30	30.46	55.11
Ireland	1981–88	22.91	27.36	36.37
Italy	1969–88	11.52	17.19	40.06
Japan	1969–88	18.07	20.99	29.35
Malaysia	1981–88	5.32	8.56	28.15
Mexico	1982–88	92.16	129.47	125.30
Netherlands	1969–88	10.98	13.31	23.85
New Zealand	1981–88	16.05	25.03	52.20
Norway	1969–88	13.99	22.21	51.45
Singapore	1970–88	20.20	29.53	53.94
South Africa	1981–88	5.64	9.18	31.42
Spain	1969–88	15.91	19.05	29.22
Sweden	1969–88	15.18	17.93	26.32
Switzerland	1969–88	4.19	6.42	22.28
United Kingdom	1969–88	13.28	18.31	37.86
United States	1969–88	9.52	10.99	17.82

Source: For 1960–80, adapted from data in *Capital International Perspective*, the Capital Group, Geneva, Switzerland, various issues. (Now updated by Morgan Stanley and made available as *Morgan Stanley Capital International Perspective*.) For 1981–88, adapted from data published in connection with the FT-Actuaries World Indices™, which are jointly compiled by The Financial Times Limited, Goldman Sachs & Co., and County NatWest/Wood Mackenzie & Co., Ltd. in conjunction with the Institute of Actuaries and the Faculty of Actuaries.

cent. The United States was below the middle of the pack. All returns are total returns (that is, they assume the reinvestment of dividends) and are before transaction costs or taxes, including dividend taxes withheld at the source.

The standard deviation of annual returns, shown in Table 2 and elsewhere, is an indication of the risk or volatility of a market. Mexican investors had the wildest ride of all, due chiefly to the stock market's response to the crashing peso. Investors in Hong Kong, New Zealand, Norway, and Singapore also faced high levels of volatility but were rewarded with high returns. The U.S. and Canadian markets were the safest in local currency terms.

Translating returns to U.S. dollar terms, Japan was an even more dramatic winner, with a compound annual return of 24.4 percent. This reflects the appreciation of the yen against the dollar over the period, as well as the performance of the Japanese market. The worst performer over 1969–1988 in U.S. dollar terms was Italy. Looking at countries with shorter data histories, South Africa was the big loser, with a compound annual return of –14 percent per year over 1981–88, at least partly caused by weakness in gold prices. Summary statistics of annual returns in U.S. dollars are shown in Table 3.

These compound annual returns do not reveal shorter-period trends, which are worth noting. Table 4 shows the compound annual return in U.S. dollars for each of 10 important countries' stock markets, where the 1969–88 period is broken into four five-year periods. In the first period, 1969–73, the world equity markets stagnated and Japan was the only really good performer.

The second period, 1974–78, showed widely dispersed results as the bear market of 1974 affected different countries to different degrees. Interestingly, the country that suffered the most in 1974, the United Kingdom, also had a very strong recovery and finished the period in the middle of the pack. Germany had the best return; Japan was second.

The third period, 1979–83, also had mixed results. The United States, which had been a laggard, led with a robust return of 19 percent per year. Japan had a modest return of 10.4 percent per year. The big loser was France, which experienced a devastating crash in 1981–82.

Over 1984–88 the worldwide bull market was led by Japan, which had an astonishing 45 percent per year return. (France had a higher annual rate of return, nearly 60 percent, but much of this return was a recovery from deeply depressed stock prices.) The returns of other countries were related to their performance in and after the price collapse of October 1987. The United States did not spring back quickly from the 1987 crash and was the second lowest returning country over the period. (The lowest was Canada.) The European markets were all strong for the first time in the period studied.

Figure 3 illustrates the growth of one unit of money invested in the four leading equity markets at the end of 1968. Returns are in the local currency of

Table 3 Summary Statistics of Annual Returns on World Equities: Total Returns in U.S. Dollars

Country	Period	Compound Annual Return (%)	Arithmetic Mean Return (%)	Standard Deviation of Returns (%)
Australia	1969–88	9.02	12.59	28.64
Austria	1969–88	11.06	14.97	37.08
Belgium	1969–88	17.45	19.50	24.05
Canada	1969–88	10.30	11.58	17.11
Denmark	1969–88	15.55	19.50	34.01
France	1969–88	13.00	22.91	53.57
Germany	1969–88	12.41	15.65	31.48
Hong Kong	1970–88	16.70	28.84	55.03
Ireland	1981–88	19.38	26.01	44.90
Italy	1969–88	7.47	14.67	47.23
Japan	1969–88	24.44	28.68	35.69
Malaysia	1981–88	2.74	5.70	26.35
Mexico	1982–88	1.29	30.11	72.64
Netherlands	1969–88	14.27	16.08	21.59
New Zealand	1981–88	10.06	19.50	54.18
Norway	1969–88	14.48	23.60	54.82
Singapore	1970–88	23.16	31.93	54.04
South Africa	1981–88	-13.99	-5.75	47.93
Spain	1969–88	13.14	17.97	36.93
Sweden	1969–88	14.22	16.64	24.09
Switzerland	1969–88	9.82	12.47	27.09
United Kingdom	1969–88	11.72	16.79	35.55
United States	1969–88	9.52	10.99	17.82

Source: For 1960–80, adapted from data in *Capital International Perspective*, the Capital Group, Geneva, Switzerland, various issues. (Now updated by Morgan Stanley and made available as *Morgan Stanley Capital International Perspective*.) For 1981–88, adapted from data published in connection with the FT-Actuaries World Indices™, which are jointly compiled by The Financial Times Limited, Goldman Sachs & Co., and County NatWest/Wood Mackenzie & Co., Ltd. in conjunction with the Institute of Actuaries and the Faculty of Actuaries.

the country indicated (i.e., in dollars, yen, pounds sterling, or deutschemarks) and include reinvestment of dividends. As suggested by the summary statistics in Table 3, Japan was the big winner. Figure 4 presents the results for the same the same countries for the same period, but returns are translated to U.S. dollars.

Table 4 Compound Annual Returns on World Equities over Five-Year Periods: Total Returns in U.S. Dollars

	Compound Annual Return over Period (%)			
Country	**1969–73**	**1974–78**	**1979–83**	**1984–88**
Australia	–0.33	4.34	14.82	18.28
Canada	11.96	1.98	17.90	9.96
France	8.15	10.37	–14.43	59.65
Germany	5.84	20.48	2.03	22.73
Italy	1.14	–11.93	7.21	39.68
Japan	27.22	17.67	10.39	45.13
Netherlands	2.34	15.50	14.06	26.49
Switzerland	7.54	15.87	1.51	15.00
United Kingdom	–0.58	10.88	13.93	24.05
United States	–0.20	5.69	18.99	14.62

Source: For 1960–80, adapted from data in *Capital International Perspective*, the Capital Group, Geneva, Switzerland, various issues. (Now updated by Morgan Stanley and made available as *Morgan Stanley Capital International Perspective*.) For 1981–88, adapted from data published in connection with the FT-Actuaries World Indices™, which are jointly compiled by The Financial Times Limited, Goldman Sachs & Co., and County NatWest/Wood Mackenzie & Co., Ltd. in conjunction with the Institute of Actuaries and the Faculty of Actuaries.

Benefits and Drawbacks of International Diversification

The question of whether or not to invest globally is addressed by two issues: gains from diversification and barriers to international investing. These barriers may be portrayed as costs of diversification. If there were no barriers, rational investors would hold variants of a portfolio corresponding to the wealth of the world. (The variations would come from differences in investors' tolerance for risk, differences in taxes, and various other clientele effects.) In this section we look first at the gains from international diversification, then at the costs.

Correlations of Market Returns

Global investing makes sense because the markets of the world do not move together. Perhaps the most dramatic example of this is October 1987, when the worldwide crash passed over Japan. The Japanese stock market had a terrible month by ordinary standards, but it did not crash, and rose to new highs within a short period. The U.S. market did not surpass its 1987 precrash high until mid-1989.

Table 5 shows the correlation coefficients of annual returns in 10 leading equity markets over 1969–87, where the returns are in local currency. None of

Figure 3 Growth of One Local Currency Unit Invested in Leading Equity Markets, 1968–88 (12/31/1968 = 1)

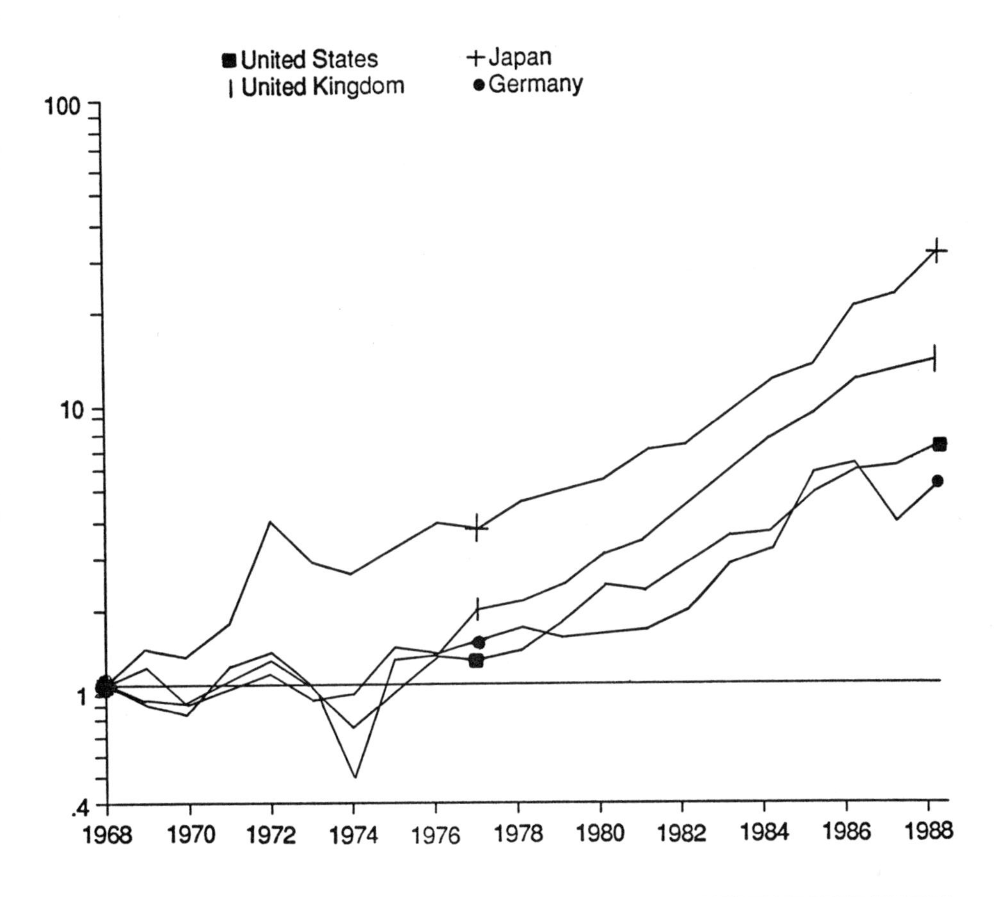

the correlations of the United States with another country is above 0.76, and the lowest is 0.34. Table 6 states the correlations of the U.S. dollar-translated annual returns of the same markets, again over 1969–88. These tables indicate the prima facie benefits of diversification. Stated simply, if assets are imperfectly correlated (which they are), there is a gain from diversification. The gain is that one can construct a global portfolio that is less risky, for the same level of expected return, than one can construct domestically.

Figure 4 **Growth of One U.S. Dollar Invested in Leading Equity Markets, 1969–88 (12/31/1968 = $1.00)**

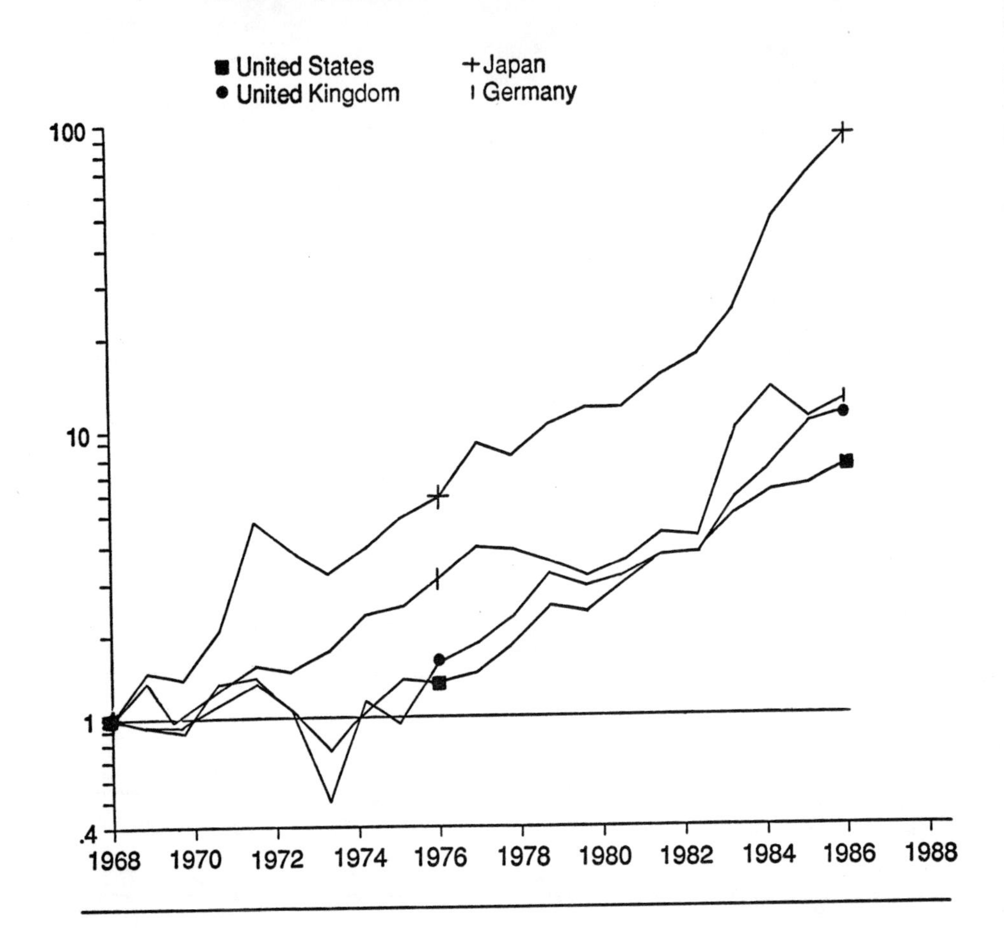

Drawbacks of Global Investing

The barriers to international investing have inhibited such investing very effectively until recently. Some of these barriers, or costs, are still a consideration in the investment equation.

Traditionally, foreign securities were a trap for U.S. investors, who often paid the highest prices and received the lowest prices (while paying extravagant brokerage commissions). Full disclosure of corporate information is an Americanism that is just now penetrating the veils of secrecy that have long shrouded foreign companies. Accounting conventions—including the most basic defini-

Table 5 Correlation Coefficients of Equity Total Returns in Local Currency, Based on Annual Data 1969–88

	1	2	3	4	5	6	7	8	9	10
1. Australia	1.00									
2. Canada	.72	1.00								
3. France	.75	.44	1.00							
4. Germany	.56	.25	.62	1.00						
5. Italy	.57	.37	.69	.48	1.00					
6. Japan	.30	.34	.38	.28	.20	1.00				
7. Netherlands	.73	.54	.55	.71	.38	.36	1.00			
8. Switzerland	.69	.55	.64	.86	.48	.41	.80	1.00		
9. United Kingdom	.52	.34	.27	.42	.07	.17	.69	.63	1.00	
10. United States	.70	.69	.40	.49	.44	.34	.74	.75	.66	1.00

Table 6 Correlation Coefficients of Equity Total Returns in U.S. Dollars, Based on Annual Data 1969–88

	1	2	3	4	5	6	7	8	9	10
1. Australia	1.00									
2. Canada	.71	1.00								
3. France	.57	.32	1.00							
4. Germany	.26	.07	.73	1.00						
5. Italy	.50	.28	.84	.63	1.00					
6. Japan	.41	.34	.49	.32	.45	1.00				
7. Netherlands	.62	.49	.75	.74	.61	.43	1.00			
8. Switzerland	.40	.30	.77	.91	.64	.45	.84	1.00		
9. United Kingdom	.58	.38	.37	.34	.25	.24	.64	.56	1.00	
10. United States	.61	.66	.35	.39	.34	.39	.76	.51	.66	1.00

tions, such as that of earnings—differ widely among countries. Foreign stocks are often subject to dividend taxation that is not imposed on most institutional investors in the United States.

Thus, transaction, information, and tax costs have traditionally been high. These barriers are eroding, but they still exist. Any cost is a decrement to return and must be considered by the investor.

Is the United States Diversified?

Another reason for not investing internationally is that (it is said) sufficient opportunities for diversification exist within the United States. Unlike some small countries where one must invest abroad in order to get a wide selection of industries and companies, U.S. investors have a rich opportunity set domestically.

We question the validity of this position in view of long-term economic trends. The world is moving towards functioning as a single economy, where each region or nation specializes in its area of comparative advantage (as states do in the United States.) The United States is increasingly a high-technology economy, where high technology includes finance, agriculture, education, and health care, as well as the more conventional high-tech fields of computer and missiles. Basic industries are making a minor comeback in the United States, but the much more important trend is for them to migrate to lower-wage areas. Extractive industries will go where the resources are.

Thus, while the United States is still fairly well diversified, it will become less so in the future. Investors should react appropriately to this long-term trend by holding a diversified international portfolio of equities.

Summary of Diversification and Barrier Issues

The question thus reduces to this: Do the gains from diversification outweigh the drawbacks? We believe that they do. The returns of actual funds invested globally are the principal piece of evidence. Since investors in these funds have paid the costs and taken on the risks cataloged above, the results already control for the drawbacks of international investment. And the returns of the globally diversified funds have been superior to, and the risk no greater than, domestic investments.

Table 7 indicates the returns on diversified and undiversified portfolios. While the highest returns would have been earned by *not* diversifying—if one knew in advance that Japan would do well—the best policy from a risk/return perspective was to hold the world equity market portfolio.

Currency Risk: To Hedge or Not to Hedge?

Hedging currency risk is a hot topic in international equity management, largely due to the volatility of the dollar over the recent past. There are two general schools of thought about hedging: (1) you should hedge, because currency fluc-

Table 7 Summary Statistics of U.S. Dollar-Denominated Annual Returns on Countries and Portfolios, 1969–88

	Geometric Mean (%)	Arithmetic Mean (%)	Standard Deviation (%)
Single Countries			
United States	9.52	10.99	17.82
Japan	18.07	20.99	29.35
United Kingdom	13.28	18.31	37.86
Portfolios of Countries			
Europe	11.70	13.77	23.03
World (ex. U.S.)	15.77	17.77	22.19
World	12.40	13.96	18.63

tuations are a risk but have no payoff; and (2) you should not hedge, because the true value of a stock is currency-independent and hedging only costs money. We look at both sides.

Dollar-denominated returns on major equity markets are compared with local-currency returns over 1981–88 in Table 8. We use monthly data. The year 1981 is used as the starting point because of constraints on monthly data, but this recent period is also a reasonable one because it captures the era of heavy crossborder investing and avoids periods of fixed exchange rates. Local currency returns are a *very approximate* measure of the dollar returns that would be achieved with a hedged portfolio; only a perfect hedge would allow the U.S. investor to receive the foreign country's local returns and existing hedges are imperfect.

With that caveat, let us look at the data. The dollar depreciated against the yen and Swiss franc over the period, making unhedged returns in Japan and Switzerland much more attractive than hedged returns. Conversely, since the dollar appreciated against the Australian dollar, French franc, Italian lira, and British pound over the period, investments in those countries had higher returns if hedged. The risk of the investment was, however, decreased in almost every case by hedging.

Interpreted naively, Table 8 supports the case for hedging. While it is difficult to predict return (or the exchange-rate component of a foreign country's equity return), less risk is preferable to more.

The danger in this interpretation is that the data period is too short to draw a firm conclusion. In most subperiods from the dollars lofty peak in February

Table 8 The Effect of Hedging: U.S. Dollar-Denominated and Local Currency Returns in Principal Foreign Equity Markets, 1981–88 (monthly data stated in annualized terms in percent)

	Compound Returns		Standard Deviations	
Country	In U.S. dollars, unhedged (%)	In local currency (a proxy for hedged) (%)	In U.S. dollars, unhedged (%)	In local currency (a proxy for hedged) (%)
Australia	9.83	14.37	30.20	24.73
Canada	8.85	8.83	20.89	18.88
France	17.93	22.34	40.74	38.56
Germany	16.61	15.17	21.82	19.69
Italy	17.36	22.45	29.06	28.06
Japan	31.99	24.23	22.48	16.94
Netherlands	21.20	20.31	19.43	19.47
Switzerland	9.58	7.40	19.85	17.22
United Kingdom	16.20	20.29	22.34	18.74

1985 to a subsequent low point, unhedged investments dominated hedged ones in both risk and return. It was profitable to get out of the dollar. We do not know which economic environment will be repeated in the future—one in which hedging pays or one in which hedging is worse than useless.

The case for not hedging rests on the branch of classical economic theory that is concerned with international parity. The basic parity theorem states that the same good should cost the same amount of money in two different places (after accounting for the cost of transporting the good from one place to the other). While some goods (say, apples) are heavy and expensive to transport, financial assets are cheap to move. Pressing a button on a computer transfers a billion dollars in cash or stock from New York to London, Singapore, or Riyadh.

In such an environment, we would expect parity to hold up rather well. When parity conditions hold, currency (money) is a *veil* that has no relevance. The investor only wants to increase the amount of goods he or she can receive in exchange for financial assets held.

If currency fluctuations are thus irrelevant, one would not bother to hedge them away. (The bother is in paying the cost of the hedging instrument or strategy.) One policy that is neutral with respect to holding an opinion on the virtues of hedging is *currency insurance*. Currency insurance is a dynamic trading strategy that attempts to take advantage of currency movements in one's favor but avoid adverse movements. It is implemented in much the same way as portfolio insurance—by trading in futures or forward markets in such a way as to replicate holding a long-term put option on the currency in question.

Conclusions

The benefits of international investing are will known. A global portfolio is more diversified than a purely domestic one; by that criterion alone, one should invest across countries. One would have to be a very strong bettor on the superiority of one's home country capital markets to be a solely domestic investor in the 1990s.

It is difficult to forecast returns on any asset, including international equities. Japan was the big winner over the last generation. This performance is unlikely to be repeatable. Possibly some country is the next Japan, but we cannot say which. It pays to hold a diversified portfolio in the hope that some of the shares will be in that country.

The benefits of hedging are far less clear. Hedging reduced risk in the 1980s, but tended to dampen return also. We anticipate that hedged, unhedged, and currency-insured portfolios will all have a place in the international investment world of the future.

CHAPTER 3

Managing International Equity Passive Portfolios

Heydon Traub
State Street Bank and Trust Company

As more and more investors allocate money overseas, many of them use passive portfolios. This style of management offers many advantages to both novice and veteran international investors.

The decision to invest outside of their home country probably results from an asset allocation review. This study clearly indicates the diversification benefits from placing money abroad, while also providing the potential for increased expected returns. Most likely, such investors have used an international index to represent expected returns and risks within the asset allocation model. Thus, logic dictates that they implement the investment with an index fund in order to realize the expected risk and return characteristics that they used in the study. The average active international portfolio provides a significantly different return stream from that of an international index. Using the three-year period ending the third quarter of 1989, the correlation of the Intersec median active manager's returns with those of the S&P 500 is approximately 0.74. If we use the MSCI EAFE Index (Morgan Stanley Capital International Europe, Australia, and Far East Index) in place of the median manager, the correlation drops to 0.47.[1] This probably results from the large shift of assets out of Japan and into Europe for

[1] Source: State Street Bank and Trust Company.

many active managers. As can be seen in Table 1, the correlation of U.S. stock returns with Japan's stock returns has been 0.24 over the last 10 years. Over the same period, most European stock market returns have had significantly higher correlations with U.S. returns. The largest European stock market, the United Kingdom, had a correlation with the United States of 0.56 over the same period. Thus, within an asset allocation model, international investments appear very attractive in diversifying risk, but by implementing it through the average active manager, investors will not realize the expected level of diversification.

A second and important advantage that index funds offer is significantly lower transactions costs. These are summarized in Table 2, and we will explain the numbers below. The average percent change in the composition of the EAFE index (the assumed international index from here on unless otherwise noted) is about 3 percent. This translates to turnover for an index fund of 3 percent. The average transaction cost for a traditional active manager trade (i.e., not a program trade) probably averages around 50 basis points, as compared to 25 for an EAFE program trade. The weighted-average tax to trade (Australia, Germany, Hong Kong, Italy, Japan, Singapore, Sweden, Switzerland, and the U.K. tax trades) is about 15 basis points. Ignoring contributions and withdrawals, and assuming 50 percent turnover each way for an active manager, this causes a drag of 65 basis points for an active portfolio, as opposed to 2 basis points for a passive portfolio. Bid/ask spreads typically represent the largest part of the transaction cost, yet managers often overlook them when moving from one security to another, because we cannot easily measure them. As with other costs, they are typically higher overseas. Based on data compiled at State Street Bank's Asset Management Division, we estimate the weighted-average bid/ask is at least 1.2 percent which translates to a cost of 60 basis points for an active portfolio (again using 50 percent turnover), as opposed to 4 basis points for a passive fund. Realistically, we would expect an active manager to face higher bid/ask spreads, since many active managers equally weight the holdings in their portfolios. This means that a larger proportion of their trades will involve smaller capitalization stocks (as compared to an index fund) that have above average bid/ask spreads.

The last component of transaction cost is market impact. This is probably the most difficult part to measure. Because of the difficulty involved, we will not add this to the transaction cost calculation. However, based on our experience managing both passive and active portfolios, market impact exists, and it is typically greater for active portfolios. This is because active managers often take large positions relative to the size (capitalization) of a company. Since capitalization correlates highly with trading volume, we should expect smaller positions offered at a given price for small cap stocks. This means that the offer price is more likely to be "impacted" up for a small stock if a manager tries to accumulate a sizable position in a short time. And we would expect that most active managers would not be willing to wait very long for fear that other investors

Table 1 Correlation Matrix for Period from 8001 to 8912 (Market Returns in U.S. dollars)

CNTRY	AUD 1	ATS 2	BEF 3	CAD 4		FIM 6	FRF 7	DEM 8	HKD 9	ITL 10	JPY 11	MXP 12	NLG 13	NZD 14	NOK 15	SGD 16	ESP 17	SEK 18	CHF 19	UKS 20	USD 21	USB 22	EAF 23
1	1.00	0.16	0.25	0.60	0.26	0.18	0.34	0.29	0.44	0.23	0.26	0.00	0.40	0.43	0.49	0.47	0.36	0.39	0.39	0.55	0.44	-0.03	0.49
2	0.16	1.00	0.38	0.19	0.17	0.12	0.49	0.58	0.17	0.27	0.13	0.06	0.40	0.13	0.31	0.04	0.28	0.26	0.53	0.29	0.14	0.11	0.32
3	0.25	0.38	1.00	0.35	0.39	0.17	0.69	0.61	0.26	0.42	0.45	-0.02	0.60	0.07	0.58	0.23	0.41	0.35	0.63	0.52	0.38	0.22	0.65
4	0.60	0.19	0.35	1.00	0.33	0.09	0.40	0.33	0.37	0.32	0.27	0.02	0.60	0.22	0.51	0.45	0.33	0.38	0.48	0.63	0.72	0.19	0.52
5	0.26	0.17	0.39	0.33	1.00	0.14	0.39	0.41	0.19	0.32	0.33	0.20	0.45	0.02	0.39	0.24	0.27	0.25	0.51	0.36	0.34	0.26	0.47
6	0.18	0.12	0.17	0.09	0.14	1.00	0.26	0.27	0.19	0.22	0.20	-0.04	0.25	0.23	0.27	0.06	0.30	0.30	0.19	0.22	0.08	0.07	0.28
7	0.34	0.49	0.69	0.40	0.39	0.26	1.00	0.64	0.23	0.49	0.40	0.01	0.58	0.16	0.54	0.16	0.40	0.36	0.63	0.53	0.42	0.19	0.63
8	0.29	0.58	0.61	0.33	0.41	0.27	0.64	1.00	0.27	0.37	0.34	0.08	0.66	0.21	0.49	0.15	0.35	0.39	0.78	0.44	0.36	0.26	0.59
9	0.44	0.17	0.26	0.37	0.19	0.19	0.23	0.27	1.00	0.31	0.22	-0.17	0.47	0.28	0.42	0.45	0.30	0.35	0.32	0.48	0.33	0.02	0.44
10	0.23	0.27	0.42	0.32	0.32	0.22	0.49	0.37	0.31	1.00	0.41	0.09	0.41	0.16	0.24	0.14	0.40	0.36	0.39	0.40	0.24	0.11	0.54
11	0.26	0.13	0.45	0.27	0.33	0.20	0.40	0.34	0.22	0.41	1.00	-0.10	0.40	0.16	0.30	0.17	0.40	0.29	0.37	0.39	0.24	0.11	0.88
12	0.00	0.06	-0.02	0.02	0.20	-0.04	0.01	0.08	-0.17	0.09	-0.10	1.00	0.04	0.01	-0.01	0.00	-0.01	0.05	0.22	-0.01	-0.04	0.12	-0.07
13	0.40	0.40	0.60	0.60	0.45	0.25	0.58	0.66	0.47	0.41	0.40	0.04	1.00	0.23	0.66	0.36	0.36	0.45	0.70	0.69	0.61	0.26	0.69
14	0.43	0.13	0.07	0.22	0.02	0.23	0.16	0.21	0.28	0.16	0.16	0.01	0.23	1.00	0.26	0.15	0.25	0.29	0.28	0.28	0.24	-0.05	0.27
15	0.49	0.31	0.58	0.51	0.39	0.27	0.54	0.49	0.42	0.24	0.30	-0.01	0.66	0.26	1.00	0.43	0.30	0.48	0.57	0.58	0.51	0.13	0.55
16	0.47	0.04	0.23	0.45	0.24	0.06	0.16	0.15	0.45	0.14	0.17	0.00	0.36	0.15	0.43	1.00	0.16	0.33	0.24	0.45	0.51	0.05	0.34
17	0.36	0.28	0.41	0.33	0.27	0.30	0.40	0.35	0.30	0.40	0.40	-0.01	0.36	0.25	0.30	0.16	1.00	0.31	0.38	0.41	0.27	0.08	0.52
18	0.39	0.26	0.35	0.38	0.25	0.30	0.36	0.39	0.35	0.36	0.29	0.05	0.45	0.29	0.48	0.33	0.31	1.00	0.48	0.49	0.40	0.19	0.49
19	0.39	0.53	0.63	0.48	0.51	0.19	0.63	0.78	0.32	0.39	0.37	0.22	0.70	0.28	0.57	0.24	0.38	0.48	1.00	0.56	0.49	0.36	0.64
20	0.55	0.29	0.52	0.63	0.36	0.22	0.53	0.44	0.48	0.40	0.39	-0.01	0.69	0.28	0.58	0.45	0.41	0.49	0.56	1.00	0.56	0.17	0.71
21	0.44	0.14	0.38	0.72	0.34	0.08	0.42	0.36	0.33	0.24	0.24	-0.04	0.61	0.24	0.51	0.51	0.27	0.40	0.49	0.56	1.00	0.27	0.46
22	-0.03	0.11	0.22	0.19	0.26	0.07	0.19	0.26	0.02	0.11	0.11	0.12	0.26	-0.05	0.13	0.05	0.08	0.19	0.36	0.17	0.27	1.00	0.19
23	0.49	0.32	0.65	0.52	0.47	0.28	0.63	0.59	0.44	0.54	0.88	-0.07	0.69	0.27	0.55	0.34	0.52	0.49	0.64	0.71	0.46	0.19	1.00
ANN %SD	30.2	24.5	23.3	21.7	20.7	31.2	25.5	23.4	36.9	27.8	22.3	*****	20.0	36.5	29.3	27.8	23.5	23.2	20.0	23.1	16.2	8.1	17.5
ANN %RET	13.5	16.7	24.3	11.4	20.1	5.2	18.3	17.4	13.9	22.6	28.6	-94.0	21.2	9.8	12.6	13.0	20.7	29.4	11.8	19.4	16.9	12.1	22.1

Source: State Street Bank and Trust Company, Asset Management Division.

will discover the "bullish" information (which has led them to want the stock) before they have bought all the desired shares. However, even assuming no market impact costs, a moderately active portfolio must select stocks that outperform by 1.19 percent in order to just match a passive portfolio's returns.

If we measure returns after management and custody fees, the drag on an active portfolio's return increases. If we assume 75 basis points for an active management fee and custody, and 25 for a passive management fee and custody, this moves the net difference to 1.69 percent. Given the new "unbundled" custody fees, each trade costs the portfolio a ticket charge. However, since custody providers charge a fixed fee, this becomes a small percentage with large portfolios (about a basis point for an actively managed $100 million portfolio). Thus, investors selecting a traditional active manager must expect that they will select stocks that will outperform the index by at least 1.69 percent per year.

A final advantage that index funds offer is the flexibility the strategy provides to add excess returns. The most straight-forward method to add return is through securities lending. In Japan, lending fees in certain situations could add 250 to 300 basis points to a stock's return, according to State Street Bank's securities lending department (although we would expect this number to fall as more players enter this lucrative market). Lending in other countries is still in its infancy, but there is certainly significant opportunity to add value with little risk. Although active portfolios could lend securities, it adds a layer of complexity and potential problems that indexers do not face. Lending might impede the active manager's ability to transact on short notice because of delays in getting the stock back. On the other hand, a passive investor rarely needs to sell a stock, and even if he does, he should have little urgency to transact because it is not an information-based trade. Other possibilities to add value to an index include futures arbitrage, options/warrant arbitrage, and index tilts such as eliminating the most unattractive stocks from an indexed portfolio.

Table 2 Annual Cost Summary—Active versus Passive*

	Active	Passive	Difference
Commissions	50	1.5	48.5
Taxes	15	0.9	14.1
Bid/Ask Spreads	60	4	56
Management/Custody Fee	75	25	50
Total	200	31	169

* In basis points (totals are rounded).

Table 3 EAFE Index Performance versus Intersec Median

Year	EAFE (%)	Intersec Median (%)	Difference (%)
1980	23.5	28.8	–5.3
1981	–1.6	–1.9	0.3
1982	–1.2	3.9	–5.1
1983	24.2	28.7	–4.5
1984	7.6	–2.9	10.5
1985	56.2	55.9	0.3
1986	69.4	60.0	9.4
1987	24.6	11.3	13.3
1988	28.3	16.4	11.9
1989	10.5	21.6	–11.1
Cumulative	650.0	544.70	132.1
Annualized	22.3	20.5	1.8

Historical data indicates that active international managers have not added value (see Table 3). Using the Intersec median manager returns, it appears that active managers have actually realized negative excess returns, even before netting out the negative effect of fees. Although this sample covers only ten years, it makes a convincing case for passive management. Combined with the more statistically significant data on active managers in the United States, it is unlikely that this period is an aberration from the norm. In fact, it is well documented that the median active manager investing in the United States has underperformed the S&P 500 benchmark over the long-run (see Table 4) by over 1 percent.

We can apply the same explanations for this underperformance in the United States, but to a greater magnitude overseas. We have already discussed transaction costs, which exceed those incurred in the U.S. market. Another factor is the difficulty managers face in analyzing all the macroeconomic data and its effects on a country's stocks. An international manager must follow approximately 18 different economies simultaneously and make fast decisions based on all this new data. Even more monumental is the individual company news. It is unreasonable to expect that one person can follow information on 1,000 companies across 18 countries, overcoming different languages, cultures, and accounting standards (not to mention time zones). Because of these huge problems, it is likely that an active international manager will have to rely heavily on outside research that in all likelihood is used by many of his competitors. Even if com-

Table 4 S&P 500 Index Performance versus SEI Median

Year	S&P 500 (%)	SEI Median (%)	Difference (%)
1980	32.57	30.60	1.97
1981	–5.34	–2.20	–3.14
1982	21.08	22.40	–1.32
1983	22.39	19.60	2.79
1984	6.11	1.50	4.61
1985	31.73	30.00	1.73
1986	18.55	16.70	1.85
1987	5.23	4.00	1.23
1988	16.83	16.80	0.03
1989	31.52	27.30	4.22
Cumulative	398.26	345.22	53.04
Annualized	17.42	16.11	1.31

petitors do not get the same information, the manager still must determine if the information is already reflected in the stock price. Finally, if the international manager decides that the company is undervalued he or she must decide if it is less undervalued than alternative stocks within and outside the country, which in most cases will require comparison of companies using different accounting methods. Thus, it should not surprise people to see that not only do active managers underperform the market, but they may actually underperform by more than the transaction costs involved. One way to overcome the information analysis problem is to use quantitative analysis. This allows a computer to simplify the problem so that the amount of information does not overwhelm the human brain. You will read more about this in future chapters.

Now that we have discussed the merits of passive international investing and, we hope convinced readers of its advantages, we will discuss the mechanics of implementation. Passive managers' greatest decision is not which stocks are most attractive, but which tracking methodology makes sense for a given index and portfolio. Full replication, optimization, and sampling represent the typical methods to choose from.

Full replication presents the most straightforward strategy. Essentially, a manager using this method would buy all the stocks in the index in the exact weightings as they have in the underlying index. This method's clearest advantage is it should have the smallest amount of tracking error (defined as the portfolio return less the index return; a passive manager's goal is to have tracking error equal to zero). This method will have the least trading involved and, therefore, the smallest annual transaction costs. The only trading required is the reinvestment of proceeds from dividends, restructurings, and takeovers, or from changes in the index (which frequently occur from takeovers or bankruptcies). This method becomes more appropriate as the number of stocks in the index falls or the dollar value of the portfolio rises. For example, it is not reasonable to attempt this method if a manager had only $5 million to invest for a separately managed account tracking the EAFE index (which comprises over 1,000 stocks).

Sampling comes next in order of simplicity. This could actually involve an infinite number of permutations. A simplistic method of this might be to eliminate the smallest capitalization stocks (say the bottom half) and reallocate their weight on a proportional basis to the top half. However, this would cause a large cap bias to the portfolio. To correct for this, a manager might choose to eliminate the same number of stocks, but eliminate them from the middle (the barbell approach!). Another option might be to randomly eliminate stocks to get down to a more manageable number of stocks in the portfolio and then transfer an eliminated stock's target weight to a similar stock in the same country and industry that is already in the portfolio (i.e., replace Nissan Motors with Toyota). The last example actually uses intuition instead of statistics, but it will approach the results of an optimizer.

Optimization is the most complicated of the three methods and provides the most flexibility. However, it requires large amounts of data and computer power. Optimization models generally focus on historical returns, and thus they require that return patterns will not change significantly in the future. One method uses quadratic programming, which given historical correlations and standard deviations (of returns) can crunch away, given certain constraints, until it finds the portfolio with the smallest predicted tracking error. A second method focuses on factors, meaning a given stock's sensitivity to certain macroeconomic occurrences (BARRA is the best known provider of these factors). A manager might present a constraint such as to limit the portfolio tracking EAFE to 300 stocks. Optimization provides great flexibility provided that a manager can get accurate inputs, i.e., the historical returns are accurate or the factor sensitivity calculated is reasonable.

Which method is best? Many of our clients have asked that question, and as with any complex question, it depends on the situation. If it truly is a passive portfolio (meaning it does not involve futures arbitrage or asset allocation models, which require frequent liquidations and contributions), we believe full replication makes the most sense. There are two goals to managing truly passive portfolios: (1) minimize tracking error and (2) minimize transaction costs (the goals are somewhat compatible, i.e., improving one may also improve the other). Full replication certainly minimizes tracking error better than the other methods, and on an ongoing basis, it minimizes transaction costs because it requires the least rebalancing. Proponents of the other methods argue that it costs more to put a passive portfolio in place, because it costs more to buy 1,000 names than, say, 400 names. This assumes that they can identify and eliminate those stocks that involve the highest transaction cost. In general, these are the smaller capitalization stocks; eliminating these would cause the portfolio to have a high capitalization bias. In addition, if an investor put money into a commingled fund tracking EAFE, it is very unlikely that the fund would buy 1,000 stocks; it would buy only those stocks it needs to fill in its positions (an index fund is never perfectly weighted because of round-lots and corporate actions, which we will discuss in more detail later). Based on an unpublished paper by Larry Martin, a Vice President of State Street Bank, even if a stratified sampling model could actually reduce transaction costs (bid/ask spreads) by buying the most liquid stocks at an average of 25 percent less (i.e., bid/ask of 90 basis points), the annual rebalancing costs would wipe out the savings within a year and a half.

Probably the strongest argument in favor of full replication is the management of S&P 500 index funds. The majority of S&P 500 index managers use full replication. Most of the large U.S. index managers have substantial assets and can easily use full replication. However, there are very few international index managers with enough assets to warrant using full replication, and as a result they champion the other methods. Thus, as long as the passive portfolio exceeds $100 million, we would favor full replication for an EAFE index.

For other indices, we would use different thresholds. For example, the FT-Actuaries Europe & Pacific Index includes about 1500 stocks, and thus requires a larger threshold of maybe $150 million. Conversely, the Salomon-Russel Europe-Pacific Index has about 500 stocks and could justify using full replication at, say, $60 million. One of the more popular indices is EAFE ex-Japan, which comprises about 750 stocks; $50 million dollars might be appropriate for this index. None of these numbers are hard and fast, but the point is that above a certain minimum size, full replication generally provides the best tracking with the smallest transaction costs.

In some sense, the version of full replication we use at State Street utilizes sampling to a small degree. That is because certain countries (Finland, Sweden, Switzerland, and Norway) have restricted shares that foreigners cannot buy, or shares have foreign ownership limits (primarily in Norway and Singapore) that may cause a two-tiered pricing system. However, in the case of restricted shares, a company has one class of shares that is restricted and one that foreigners can own (but often confer no voting rights). In these cases, we suggest buying the "free" shares (i.e., no restrictions), which in most cases track well with the restricted shares. Similarly, in those countries with foreign ownership limits, when a stock has reached its limit, an indexer could buy the "foreign-registered" shares (from another foreigner), albeit at a premium price. The most difficult restriction comes in Sweden, where foreigners cannot own any shares of banks.

This presents a problem because, ideally, a sampler would buy a stock in the same country and industry when substituting. However, the best one can do in this situation is to either replace it with stocks in the same country or banks in another country. Despite the apparent problems, the stocks involved that cannot be bought in any form are far less than 1 percent of EAFE, and even those that involve substituting one class of shares for another involve less than 4 percent of EAFE. To deal with these restrictions, some investors have decided to use EAFE-free as a benchmark, since its returns are calculated excluding the restricted stocks. In addition, the Salomon-Russell Europe—Pacific Index excludes shares that foreigners cannot buy.

There are some situations where significant use of sampling or optimization makes better sense. Beside the already mentioned small portfolio size, South Africa-free portfolios should use either sampling or optimization. Because the investor excludes a substantial number of stocks (around 300 for an EAFE portfolio), the manager should attempt to find suitable substitutions for the restricted stocks. In general, we would expect an optimizer to provide better tracking, but it clearly depends on how accurate the historical data is and whether history is a good predictor of future return patterns.

Passive managers may also use index futures (either alone or in conjunction with equities), but this also would currently require use of sampling or an optimizer. That is because (1) only a few of the EAFE countries have index futures (Australia, France, Hong Kong, Japan, New Zealand, and the United Kingdom, (2) the underlying indices do not match those in any of the major international indices discussed above, and (3) the CFTC only allows U.S. investors to trade the Nikkei 225 futures in Singapore (although traded in Singapore, it is an index of Japanese stocks) and the FT-SE 100 in the United Kingdom. It is likely that in the near future, we will have EAFE futures, although because of the difficulty in arbitraging these with the underlying index (all the stocks trade at different times), it is not clear how liquid they will be.

In addition, sampling and optimization may help in creating an enhanced portfolio or an "index plus" strategy. For example, a manager may eliminate some of the companies in the index because of poor expected returns, but through an optimizer he may still create a portfolio with tight tracking to the index and create a positive alpha. Next, we will focus on the mechanics of managing a passive international portfolio.

Managing a passive international portfolio does not differ significantly from managing a passive domestic portfolio. As in the United States, the target weights are simply a given company's market value (outstanding shares times price). Of course, since we are dealing with multiple currencies, we must then convert the market values to one currency. Next, we must deal with any unusual situations, e.g., restricted shares, which we discussed above. For example, using a full replication strategy, we must transfer a restricted stock's target weight to a substitute stock. So in Sweden we would transfer Saab-Scania's target weight to Saab-Scania Free shares. With optimization we would simply put an upper weight of zero on the restricted shares as a constraint to the optimizer. Once we have adjusted target weights, we would need to determine the target number of shares to buy, which would be the portfolio value times a stock's target weight divided by the (dollar) price per share. Finally, we would need to round to the appropriate round-lot. These last two steps are what makes full replication inappropriate for very small portfolios. What happens with the smaller companies is that we would round to zero shares and, thus, even if we wanted to use full replication, we might only be able to buy half the companies in the index due to round-lots. One complication introduced by investing in multiple countries is that countries have varying round-lots. Japan and Hong Kong generally use 1,000 whereas some companies in other countries are as little as one. In addition, unlike the United States, where 100 applies to the vast majority of companies, several countries have a variety of round-lot sizes. Lastly, certain countries (most notably Japan) will not allow odd-lot trades, whereas most countries will allow it, but it will frequently increase the transaction costs incurred.

Maintenance of a passive international portfolio does not vary much from its U.S. counterpart. The primary advantage of portfolio weights "automatically" changing to match the target weights still exists. Also, as is true in the United States, the greatest cause of change to an index is corporate actions. The most dramatic changes to an index typically involve a takeover or recapitalization. Although not yet as common as they have been in the United States, they have been occurring with increasing frequency overseas. Typically, this will involve a cash flow to a portfolio (from tendering or selling) and then reinvestment with some or all of the cash used to buy a replacement company (in many cases). Unlike the United States, rights issues occur frequently (particularly in the United Kingdom) in international portfolios. The usual response should be to take up the rights (assuming they allow you to buy new shares at a below-market price). The reason a passive manager should take them up is that the bulk of the rights holders will take them up, causing an increase in the number of shares. The index will eventually reflect the increase in shares, causing a passive portfolio to need additional shares. If investors exercise all the rights, then, ultimately, by taking up the rights a passive manager will have exactly the desired number of shares.

The last significant part of maintaining an ongoing passive portfolio involves the reinvestment of dividends. Ideally, a passive portfolio remains 100 percent invested at all times. Realistically, no portfolio is ever exactly 100 percent invested, primarily because of dividends. It is imprudent to attempt to reinvest dividends as soon as the manager receives each payment, as that would involve many extremely small transactions. Thus, the manager must use his judgment (yes there is some, albeit small, amount of discretion involved in managing a passive portfolio) as to when to reinvest the accumulated dividends. With the EAFE index, the dividend yield is only 1.5 percent, and thus it is less of an issue than for a U.S. passive portfolio with a yield of about 3.3 percent (as of the end of 1989).

As we have shown above, passive international portfolios have inherent advantages over their active counterparts. Clearly, the most compelling case comes from the lower transaction costs involved, which makes the active portfolio manager's task even greater overseas. Although the passive manager has few traditional investment decisions to make, the two that an investor must make up front involve the index to track and the method to do it. We discussed briefly the differences in the various international indices. Although each is distinct, the difference in historical returns among the three is small. We believe the method of tracking merits more consideration than the choice of index, although, admittedly, the difference in returns will also be small. Under the majority of circumstances, we believe full replication offers the best of both worlds: lower transaction costs and, without a doubt, better tracking. Last, although management of purely passive portfolios requires no stock-picking ability, it is not a "no-brainer" as some claim. Passive management requires knowledge regarding the maintenance of an index, understanding of how stocks are traded (particularly important for the initial investment of funds), and investment savvy to deal with the occasional difficult situations that can arise with various corporate actions. The huge amount of assets committed to passive investment in recent years indicates that more investors see the merit of at least some passive international investment. Clearly, there is room for sound enhanced or active strategies as well, and you will hear about some of these in future chapters.

CHAPTER 4

International Portfolio Diversification with Stock Index Derivatives

Robert G. Tompkins
Minerva

It is a well documented fact that international diversification can yield a superior risk/reward trade-off compared to domestic diversification. The problem with international diversification is that the strategy can be difficult to implement and has numerous additional risks, including currency exposures, investment selection difficulties, and problems with regulations and capital restrictions for foreign investors. In this chapter we will review the problems inherent with international equity investment and propose new strategies for international portfolio diversification that have recently become available as a result of the introduction of futures and options on stock-market indices around the world. These strategies offer the investor a natural hedge against the most commonly identified problems associated with international diversification.

Problems with International Investment

With the double benefits of higher expected return and lower expected risk, it would seem that those investors not choosing to invest internationally would be at a disadvantage. However, this may be an oversimplification when one considers the obstacles that have until recently existed for international investment.

Investing abroad introduces risks, including currency risk, political or sovereign risk, liquidity risk in foreign markets, and capital restrictions on expatriation of profits (all of which are not specified in most asset-pricing models). Difficulties also exist in choosing which assets in each country make up the best portfolio mix. These risks complicate the use of portfolio techniques such as CAPM (Capital Asset Pricing Model) or its more cosmopolitan relative IAPM (International Asset Pricing Model). These models, which assume the separation of investment risk into systematic and firm-specific risk, will be misspecified if such factors such as currency movements are systematically related to the returns of these firms. Therefore, the international investor must carefully examine the effects of foreign exchange on international diversification.

Let us assume that the returns to an international investor can be gained from two separate sources of return—the return of the asset in terms of the local currency and the appreciation or depreciation of the asset currency versus the investor's home currency. To simplify matters, let us also assume that the investor is ultimately based in one currency such as U.S. dollars. The local returns and the impact of currency moves for a U.S. investor for the period from 1970 to 1986 can be seen in Figure 1. In this chart, the impact of currency returns to the total return of the U.S. investor seems to be unsystematic. That is, for some markets such as Japan, Germany, and other non-U.S. equity markets the effect is positive, while for the United Kingdom the effect is negative over this period. However, we cannot simply restrict our examination to the realized returns but must closely review the contribution of currency movements to the total risk of the portfolios. Figure 2 displays the currency contribution to total risk, and this risk is significant and systematically positive. When we break down the total risk of international equity investment into the local return and the currency impacts, we can see that the currency risk is approximately equal to 50 percent of the pure equity risk of these markets. In Figure 2, the total risk for each country is represented by the outlined bar graphs, with the pure equity and currency risks represented by the smaller bars within. Let us consider Japanese equity investments for U.S. investors, the total risk of Japanese investments was around 22.75 percent during the period from 1970 to 1986. The equity risk alone was 19.67 percent, and the currency impact was around 10.3 percent. Investing in foreign equity could be considered equivalent to a portfolio of pure equity investment and a pure currency asset. In the Japanese risk column, the correlation coefficient between the equity and currency changes is .05. For the U.K. equity investment, the risk of the equity alone was 24.5 percent for the period, with a currency risk of 10.4 percent. Finally, for Germany the total risk of investment is 19.7 percent for this period, with a pure equity risk of 15.9 percent and a currency risk of 11.4 percent. While we have not reviewed the breakdowns for all foreign equity investments, we will assume that the international investor will face considerable currency risk whenever investing in foreign equi-

Figure 1 International Equity Markets Local and Currency Returns, 1970–86

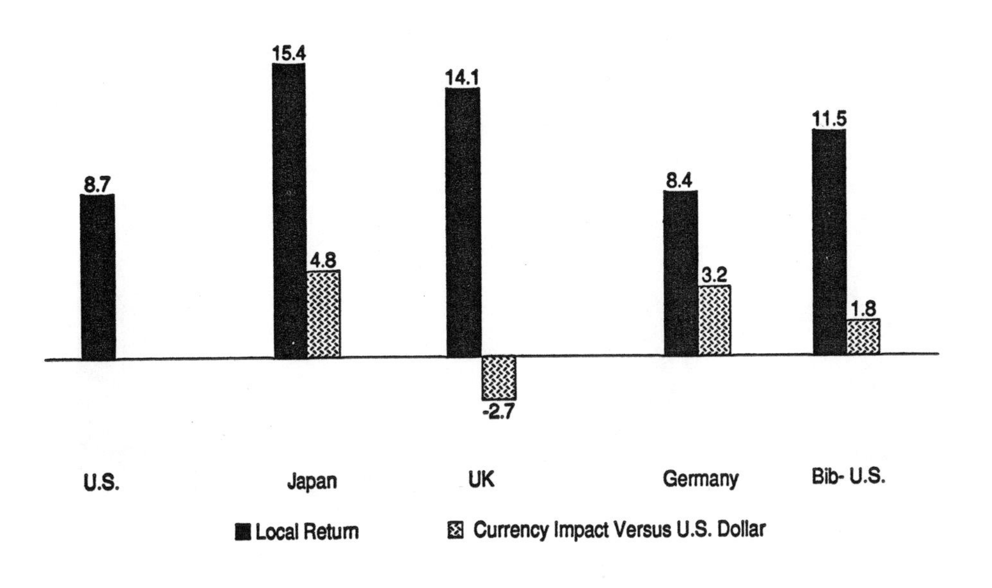

Source: Adrian Lee, "International Asset and Currency Allocation." *In The Handbook of International Investing*, ed. C. Beidleman (Chicago: Probus Publishing, 1987, p. 309.)

ties. Therefore, the equity investor who decides to invest internationally must recognize and manage the currency exposures. For many portfolio managers, the foreign exchange market is unfamiliar, with many unforeseen hazards and complications.

To manage currency risks, the most obvious hedging technique is to borrow the exact amount of the intended foreign equity investment in the appropriate foreign investment and pay off the loan when the shares are sold. Another alternative would be to use the forward foreign exchange markets to repatriate investment proceeds back to the home currency. The expected cost of borrowing abroad will be the difference in interest between the home and foreign country, which should be offset by the expected percentage change in the exchange rate.

Figure 2 International Equity Markets Local and Currency Risks, 1970–86

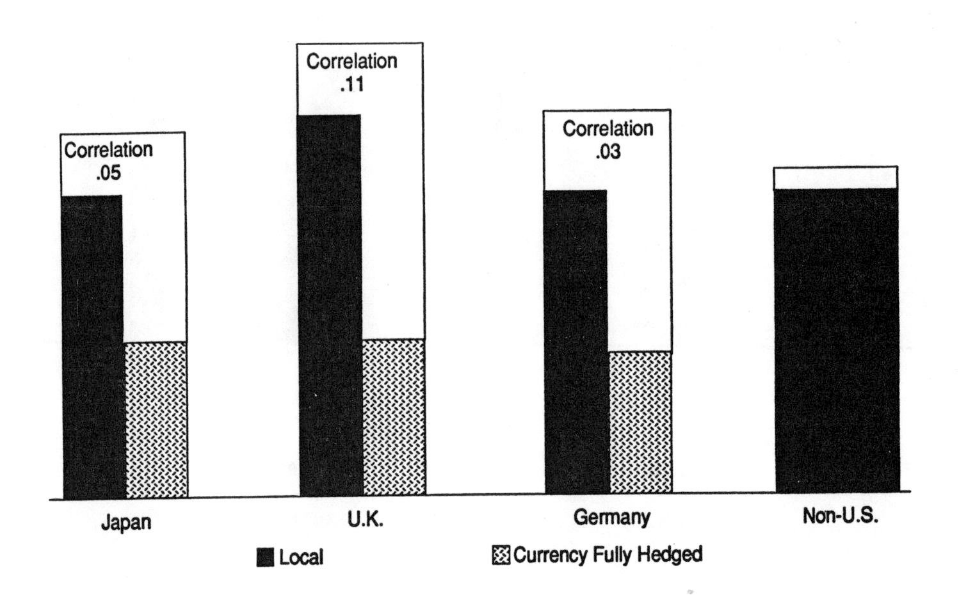

Source: Adrian Lee, "International Asset and Currency Allocation." *In The Handbook of International Investing*, ed. C. Beidleman (Chicago: Probus Publishing, 1987, p. 311.)

The forward contract hedge will yield a similar cost (because the forward rate is also determined by the difference in interest rates and is commonly referred to as the Interest Rate Parity Theorem (see Aliber 1973). This hedging can be expensive, but empirical studies have shown the cost to range between 0.5 to 0.7 percent per year for most major currencies (see Lessard 1988). A significant problem with this technique is that it assumes the date of the asset revaluation and implied liquidation (hence the paying back of the borrowing or settlement of the forward contract) is known by the investor when he initially makes his investment. However, the investor often will liquidate the investment only after it has reached a certain profit objective. If buy and hold is the best policy (as is the prevailing wisdom), the currency hedge must be held in place for an indeterminate term, and this will incur ongoing costs, not to mention an additional source of variability for the international portfolio.

Another significant risk in international investment is political or sovereign risk. If markets are efficient and all available information is reflected in security market prices, the expected returns of holding assets of these countries should reflect their level of political stability. If political risks are domestic phenomena, a substantial amount of this risk could be diversified away through international investment.

A more significant problem is accessing these foreign markets and selecting the most efficient portfolios (see Fama 1976). Many of these markets are small, dwarfed by the larger and more liquid equity markets in the United States, Japan, and the United Kingdom. For example, the Australian, New Zealand, French, Swedish, and Finnish markets might have the perfect risk/return characteristics to meet a particular investor's objective in an international portfolio. However, dealing in these markets requires learning about each market, and there is no guarantee that the investment can actually be purchased because of liquidity or legal constraints. As Lessard (1976) points out, for many of these smaller markets, most of the equity-market capitalization is concentrated in a relatively small number of companies. Often, this capitalization is misleading, because most of the shares outstanding are not available for trading but are held by institutional investors in that country. Another problem is that many countries (such as Japan) limit the access foreigners have to their domestic equity markets. This limitation has led to the introduction of vehicles such as ADRs (American Depository Receipts) to satiate international investors' appetite for foreign equity investments. Nevertheless, some investors are still willing to expend considerable resources to implement an international investment scheme. While risk reduction and return enhancement seem to reward those who internationally diversify, many risks remain that are difficult to evaluate using traditional methods of portfolio theory (which assume a single currency base). Ideally, the international investor will find some strategy whereby individual asset selection is unnecessary and the restriction on capital holdings is minimized. Until now, these restrictions have confined the benefits from international diversification to the theoretical; however, this situation has changed with the introduction of cash settled futures and options on broadly based stock market indices tied to a significant number of the world's stock markets.

Stock-Index Futures: Internationally

Stock-index futures have certainly become an important tool for the U.S. portfolio manager since their introduction in 1982. These instruments are used in stock portfolio hedging, program trading techniques such as portfolio insurance, and to achieve indexation. As of February 1990, stock index futures have been intro-

Table 1 Stock-Index Derivative Instruments Available

Exchange	Stock Index Contracts
American Stock Exchange	Major market index (XMI) option Institutional index option International market index option Computer technology index option Oil index option
Chicago Board of Trade	Major market index (MMI) future Institutional market index future CBOE 250 index future
Chicago Board Options Exchange	S&P 100 option S&P 500 option CBOE 250 index future
Chicago Mercantile Exchange	S&P 500 future S&P 500 option
Coffee, Sugar, & Cocoa Exchange	International market index future
Copenhagen Stock Exchange	Danish stock-index future Danish stock-index option
Deutsche Terminbourse	DAX index future
European Options Exchange	EOE Dutch stock-index future EOE index options
Finnish Options Brokers Ltd.	FOX index future FOX index option
Hong Kong Futures Exchange	Hang Seng stock index future
Kansas City Board of Trade	Value Line stock index future Mini Value Line stock index future
London International Financial Futures Exchange	FT-SE 100 index future
London Traded Options Market	FT-SE index option
Marche a Terme International de France	CAC 40 stock-index future
New York Futures Exchange	NYSE composite stock index future Russell 2000 stock-index future Russell 3000 stock index future
New York Stock Exchange	NYSE composite stock-index future
New Zealand Futures Exchange	Barclays index future Barclays index option
Options Market of France	OMF 50 future
Osaka Securities Exchange	Nikkei 225 future Osaka 50 future (physical delivery)
Pacific Stock Exchange	Financial News Composite index option
Philadelphia Stock Exchange	National OTC stock-index future National OTC stock-index option Value line composite-index option Utility stock-index option
Singapore International Monetary Exchange (SIMEX)	Nikkei 225 future
Stockholm Options Market	OMX 30 Swedish stock-index option OMX Swedish stock-index future

(Table continues)

Table 1 Stock-Index Derivative Instruments Available (Continued)

Exchange	Stock Index Contracts
Swiss Options and Financial Futures Exchange (SOFFEX)	Swiss market-index option
Sydney Futures Exchange	All ordinaries stock-index future All ordinaries stock-index option
Tokyo Stock Exchange	Tokyo stock price index (TOPIX) future
Toronto Futures Exchange	TSE 300 composite option
Toronto Stock Exchange	TSE 35 stock-index future TSE 35 stock-index option

duced on indices for 14 of the world's stock markets. Table 1 lists the stock-index derivative instruments available.

At this point, a brief review of stock index futures is in order. Consider a stock index such as the Standard & Poor's 500 Composite Index. This index is compiled by adding up the capitalization of the 500 largest publicly traded stocks in the U.S. equity markets. Indexing works because percentage changes in the level of the index mirror the percentage change in wealth from holding a portfolio with the same composition. Therefore, the index can be seen as a substitute and equivalent portfolio to holding all the equity in that index. However, an index is not a tradable asset, necessitating the construction of tradable securities based upon the index. Thus, the development of futures and options contracts on indices fills this gap. To give these instruments a monetary value, the index level is multiplied by a constant cash amount (such as $500 times the index); this allows a change in the index level to be associated with a change in wealth. Finally, the futures or options contracts must specify some point in time at which the level of the derivative stock-index product (futures or options) converges with the level of the underlying stock index. For futures contracts, this convergence occurs on the final day of the futures contract life. (On the cash settlement date, the futures price is set exactly equal to that day's stock-index level.) Therefore, the user of a stock-index futures contract knows that at some point he will have a position that would have monetary payoffs identical to holding the underlying stock-market portfolio (ignoring dividend payments).

That makes these stock-index futures so attractive relative to the underlying stock investment is that they have very different costs. Apart from lower transactions costs, futures require much less capital to be used when establishing positions. For example, a futures on a stock index can be thought of as costless and purely levered when one considers that the establishment of a position requires only the posting of a good-faith deposit (margin) that will be returned upon the maturity of the contract. Since this margin can be met with interest-bearing securities in many cases, there is no opportunity cost loss. Because most futures contracts settle and are marked to market each day, futures cashflows parallel the movement of the cash portfolio cashflows from initiation of the futures to their final settlement. As is well documented, a cost-of-carry relationship (along with tax implications [see Cornell & French 1983]) will define the arbitrage tie between the futures and the cash market levels. Therefore, one could conceivably hold one's entire endowment in some risk-free asset and purchase stock-index futures to create a synthetic market portfolio. While dividends are not paid to the stock-index futures holder, they are reflected in the pricing of the futures contracts.

If a broadly based index such as the S&P 500 does approximate the true market portfolio, then the investor has identified a vehicle that should allow him to achieve an efficient market proxy. He can achieve the best possible risk/return payoff by combining a holding (or borrowing) in some risk-free asset and purchasing (or going long) this market proxy. This logical extension of the CAPM will not be the focus of this chapter, since this has been extensively discussed elsewhere (see Stulz 1985). What is of interest to us is the extension of this methodology internationally. The international investor could combine expected risks, expected returns, and correlation coefficients of all markets that have stock-index derivative products and define an efficient frontier for that international equity portfolio. Once this has been achieved, the investor could combine his own particular risk-free asset with this international market portfolio to achieve a new capital-market line.

A significant problem exists for the CAPM methodology if no risk-free asset exists. (Black [1972] addressed this issue with his zero beta portfolio). We are concerned that a particular investor's risk-free asset (or zero beta portfolio) becomes currency dependent as he spreads his investment internationally. Our approach using stock-index futures and options ameliorates this problem by reducing the amount of foreign currency commitment and thus minimizes the impact of currency changes compared to investing the entire endowment in the actual purchase of foreign stock. Consider a U.S.-based investor, he could conceivably hold his entire wealth in a U.S. currency risk-free asset such as U.S. Treasury zero-coupon instruments and then use these instruments as the good-faith deposit (margin) necessary to purchase or sell stock-index futures contracts on as many foreign markets as would be necessary to achieve his ideal international portfolio. Furthermore, a similar strategy could be implemented by a Japanese investor who chooses his appropriate currency risk-free asset (which would probably be a Japanese government short-term, interest-bearing instrument) and purchases or sells stock-index futures appropriate to his investment objectives. It is important to consider that the appropriate risk-free asset will depend on the investor's home country. Also, the international asset allocation mix will depend on the investor's method for selecting the composition of his portfolio.

Another benefit of futures contracts is that it is as easy to establish buying (long) positions as it is to establish selling (short) positions. Let us assume that a particular asset-allocation scheme would suggest shorting a number of the foreign stock markets and investing the proceeds by purchasing other countries' stock markets. If the investor used the actual foreign stock markets, this would be impossible to achieve because of restrictions in many of them on short selling. Even when such restrictions do not exist, it may prove difficult to short sell the entire market. However, with stock-index futures, it is a simple matter to sell the index and thereby replicate a short selling position in that particular market portfolio.

As stated previously, futures strategy reduces the currency exposure for the international portfolio manager, because the majority of the investment is held in the home currency. In addition, the investor does not have to select individual stocks in each market, since the stock indices represent a broadly based surrogate. Finally, in many countries the investor will not face the same degree of restrictions using futures, because these markets often encourage international participation to maximize the liquidity required to operate effectively. Nevertheless, the futures strategy does have potential problems.

The currency risk has been reduced, but it is not eliminated. Most futures markets require a good-faith deposit of between 5 and 10 percent of the value of the underlying contract. If these markets permit the posting of the initial margin in another currency (or allowed interest-bearing securities in that other currency to satisfy the requirement), the investor could retain all of his assets in his own currency and simply use their value to establish futures positions, and the foreign exchange risk would be reduced further. However, most markets will accept only margin based in their own currency. So the investor would be obligated to post at least 10 percent of his assets in the securities of other currencies to serve as the basis for initial margin. Another problem is that when the price of futures contracts changes (potentially on a daily basis), the clearing house of the futures exchange will require the investor to pay (in the local currency) variations to his margin account to assure performance. Of course, the futures price could move in the investor's favor, which would result in the inflow of foreign currency to his margin account. Regardless of the direction of the cashflows, the investor will find himself with a static currency position requiring monitoring and possibly revision. Nevertheless, it is highly unlikely that the currency exposure for the futures strategy will ever approach the risk of the cash-market strategy of borrowing in the foreign currency or using forward contracts on the full amount of the foreign stock-market portfolio.

Stock-Index Options: Internationally

Now, let us consider the use of options on stock indices. Table 1 also lists the available stock-index options. These derivative products share many of the positive attributes of futures on stock indices (many of the stock-index options are actually options on stock-index futures) without some of the detrimental aspects of the futures markets. How do these instruments work? Like the futures markets, options allow an investor the ability to establish a position that mimics the underlying market portfolio. One difference for options (relative to futures) is that the position is established at a fixed strike price (as opposed to futures contracts, which have their price adjusted each trading day). A more important difference is that options have an asymmetrical payoff structure. Figure 3 displays the profit/loss profile diagrams for long futures versus holding call options and short futures versus holding put options. The futures strategies present a potential problem for the international investor, because futures contracts require variations in the value of the contracts to be settled on a daily basis. Gains to futures contracts are less of a problem for the investor, because gains do not require an immediate foreign-exchange decision. Futures losses will require additional currency to be converted to pay the variation margin. For the holders of options, once they have paid their premiums (except in those markets that allow the investor to margin the premiums), no further cashflows are required prior to the expiration of the options contracts. If the option expires worthless, the investor will not assume a position in the underlying market or receive any proceeds back in the foreign currency and consequently will have no currency exposure. However, if the option expires with positive value, the investor can realize this inflow as a favorable impact for his portfolio. However, any inflows will have a foreign-exchange exposure.

Consider the following portfolio strategy. An investor could place the bulk of his investable assets in a security paying his domestic risk-free interest rate for his investment horizon. He converts his remaining assets into the relevant currencies and purchases options on the available stock indices for those foreign (and possibly domestic) equity markets of interest.

What is the range of possible outcomes for his portfolio at the expiration of the options contracts? Suppose that each and every equity market fails to move as the investor expected. Then all of the options will expire worthless, and he will have lost the premium paid for the options. However, since the bulk of the investment was placed in his home currency risk-free asset, he can determine a priori an absolute minimum value of his total holdings at the investment horizon. In high-interest-rate environments, the investor can find a mix between the risk-free asset and holding options that can guarantee the preservation of his principal regardless of the movement of the relevant stock markets. However, if any of the chosen equity markets performs as anticipated, the investor will be able to add these proceeds to his portfolio. The exact amount of the proceeds will de-

Figure 3 Comparison of Stock-Index Futures and Options

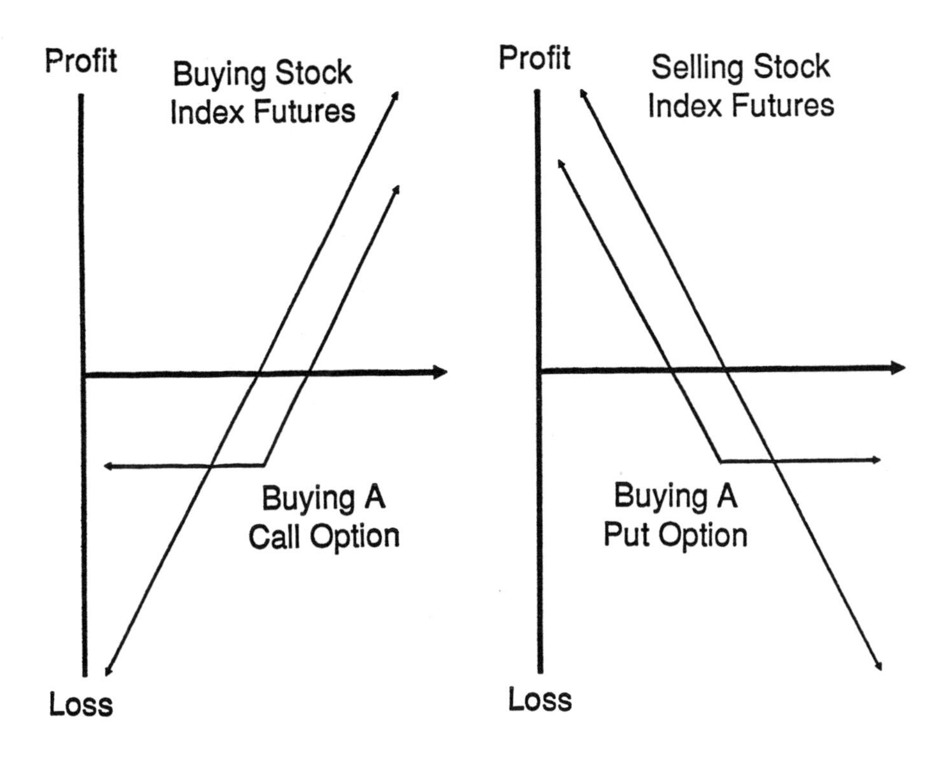

pend on both the value of each option contract and that exchange rate for the currency associated with the option. Regardless of the exchange rate, if an option has value, there will be a positive inflow to the portfolio.

Let us compare this option strategy to a 100-percent investment in an international equity portfolio using physical stock. As discussed before, when purchasing the actual stocks, almost 50 percent of the portfolio's risk will be defined by the currency exposure. For the options strategy, since only a small fraction of the total assets are invested in the other currency via stock-index options (for example 10 percent of his capital), the currency risk would be only 50 percent of this fraction (10 percent of his option invested capital would yield a maximum currency risk of 5 percent).

This investment strategy, combining options and a risk-free interest instrument, is very popular in the domestic U.S. market. The best-known variation on this theme is known as 90/10, where 90 percent of the investor's assets are placed in short-term, interest-bearing securities, and 10 percent is used to purchase call options on individual stocks. Figure 4 reviews simulations of a variety of 90/10 call-options strategies completed by Merton, Scholes, and Gladstein in 1982. For the period from 1963 to 1975, a simulation strategy of 90/10 seems to outperform simply buying and holding physical equity. Strategies using option strike prices that were at the money (E = 1.00) and those strike prices that were out of the money (E = 1.10 and E = 1.20) performed the best. The authors also found that the technique significantly reduces risk relative to holding the underlying equity. For the international investor, this type of strategy promised to reduce risk further by addressing both equity-market risk and the currency risk. Other benefits are that the need for continuous cashflows is minimized because the option premium has already been paid and that it is unnecessary to pick stocks in each equity market, since each stock index represents a well-diversified market portfolio for that particular country. Finally, these derivative products often have greater liquidity than the underlying equity market, thus facilitating implementation of large buying or selling programs.

Potential International Stock-Index-Futures/ Options Strategies

In this section, we assess which foreign stock-index markets an investor might want to participate in, how he will allocate his investment among these markets, and what possible payoffs might be obtained from some simple strategies.

The selection of which markets to participate in could be determined in a variety of ways. The easiest way to mimic an equity-buying program would be to purchase futures contracts or purchase call options on all available markets. One naive allocation scheme would be to split the assets for equity investment evenly among all the markets. In a call-options buying strategy, this is similar in spirit to betting evenly on all the horses in the Derby. As long as some of the horses finish in the money, the investor will be ahead. Of course, he can lose the entire amount he bet if each and every market fails to move up.

Another method of allocating his investment is to replicate an international stock index based upon total capitalization. Table 2 lists the capitalization for the world's major stock markets and the weighting for the World Stock Index and the EAFE (Europe, Australia, and the Far East) Index as compiled by Morgan Stanley Capital International Perspective (MSCIP). Of the 21 countries in the MSCIP World Index, 14 have either stock-index futures or options available: Australia, Canada, Denmark, Finland, France, Germany, Hong Kong, Japan, the

Figure 4 Growth of $1,000 in Four Options/Paper Strategies and Stock

A: 136-Stock Sample

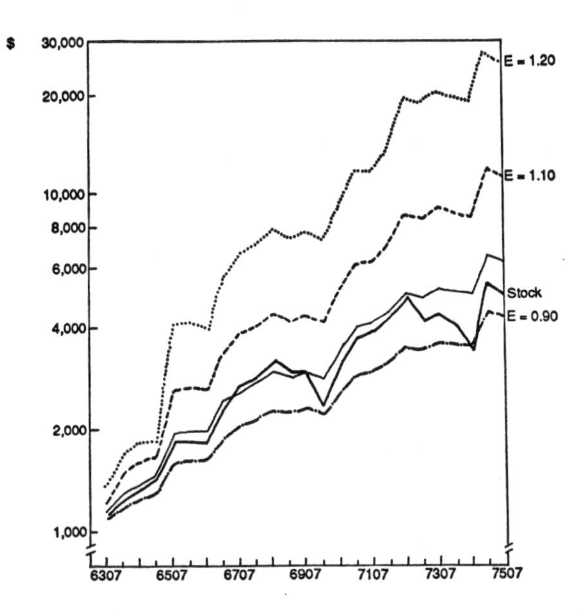

B: DJ Sample

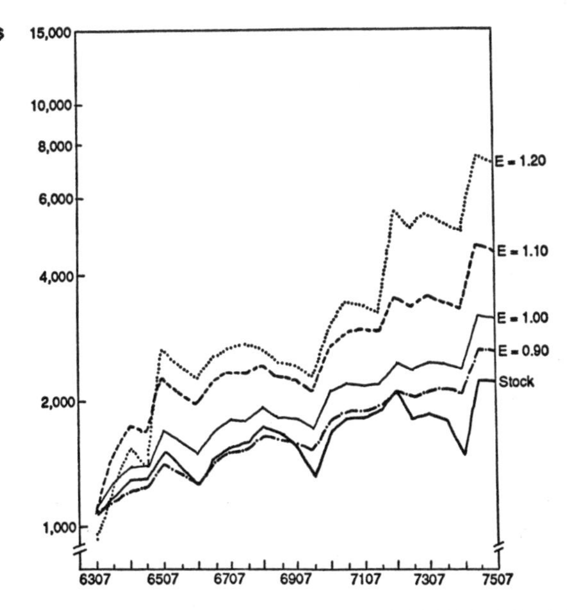

Source: Merton, Sholes & Gladstein, "The Returns and Risks of Alternative Call-Option Portfolio Investment Strategies." *Journal of Business* 1978, vol. 51, no. 2, pp. 183-243, figs. 4, 5, 7, 8.

Table 2 Capitalization of World's Major Stock Markets and Weighting for the World Stock Index and the EAFE

Country	Market Capitalization in index $ Billions	Weight as % of MSCIP World Index	Suggested "World" Allocation Weight as % of stock index derivative markets	Weight as % of MSCIP EAFE Index	Suggested "EAFE" Allocation Weight as % of stock index derivative markets
Australia	82.6	1.4	1.4	2.1	2.2
Austria	12.9	0.2	–	0.3	–
Belgium	38.9	0.7	–	1.0	–
Canada	162.2	2.7	2.8	–	–
Denmark	22.7	0.4	0.4	0.6	0.6
Finland	11.5	0.2	0.2	0.3	0.3
France	188.4	3.2	3.3	4.8	5.1
Germany	223.7	3.7	3.9	5.7	6.0
Hong Kong	41.2	0.7	0.7	1.0	1.1
Italy	93.5	1.6	–	2.4	–
Japan	2366.1	39.6	41.4	59.8	64.0
Netherlands	91.9	1.5	1.6	2.3	2.5
New Zealand	9.3	0.2	0.2	0.2	0.3
Norway	14.3	0.2	–	0.4	–
Singapore	36.5	0.6	–	0.9	–
Spain	62.9	1.1	–	1.6	–
Sweden	63.4	1.1	1.1	1.6	1.7
Switzerland	102.9	1.7	1.8	2.6	2.8
UK	494.5	8.3	8.6	12.5	13.4
US	1846.1	30.9	32.3	–	–
S. Africa Gold	15.1	0.3	0.3	–	–
Total	5980.6	100.0	100.0	100.0	100.0

Source: Morgan Stanley Capital international perspective.

Netherlands, New Zealand, Sweden, Switzerland, the United Kingdom, and the United States. Since the investor might be able to use gold futures or options as a surrogate for South African gold shares, 15 markets of the 21 countries in the World Index could be replicated using derivative products. These 15 markets constitute 95.5 percent of the capitalization of the MSCIP World Index. The remaining 4.5 percent is spread among six fairly small equity markets: Austria, Belgium, Italy, Norway, Singapore, and Spain. While one cannot ignore the exclusion of these markets, it is quite possible that many of these markets are highly correlated to one or more of the 15 markets included in our sample (for example, Norway with Sweden or Austria with West Germany). In addition, most of these markets have plans to introduce futures and options on domestic stock indices within the next few years. At that time, 100-percent coverage of the world stock index will be possible. Nevertheless, from a capitalization standpoint, 95.5 percent coverage should be adequate for the purposes of approximating the broadly based World Index.

The allocation process to replicate the World Index would entail dividing the assets available for investment and distributing them among the market by the share of worldwide equity value that each market has. Table 2 also lists the allocation percentages for each of the individual countries' stock indices. For example, 41.4 percent of the disposable investment would be placed in Nikkei-Dow stock-index futures or options to simulate investment in the Japanese stock market, and 32.3 percent of the investment would be in S&P 500 stock-index futures or options (or one of the many U.S. equity-index derivatives) to mimic a U.S. equity investment.

It would also be possible for the investor to replicate the EAFE index using individual countries' stock-index futures or options. However, only 93.5 percent of the capitalization of that index would be represented with markets that have derivative products. Table 5 also lists the weightings for the MSCIP EAFE index and the percentages we would recommend in each futures and options market to replicate that index.

Of course, the ideal situation would be the existence of a World Stock Index or EAFE Index futures and options. In fact, a "world" index on 50 international shares exists at the Coffee, Sugar, and Cocoa Exchange. It may seem to make little sense for the portfolio manager to replicate a World Stock Index if such an index already trades in a futures and options market. However, the CSCE contracts (which are based on ADRs) often fail to follow World Stock Market indices, and many potential customers have been dissatisfied with the performance of these instruments. Therefore, selecting an optimal mix of individual stock-market-index futures and options in an overall portfolio scheme could still provide the investor with more consistent replication and would allow the investor more opportunities for complicated allocation strategies.

Suppose an investor chooses those markets in which stock-index derivative products are offered and applies a Markowitz minimum variance portfolio scheme to determine the efficient set. By including expected returns, variances, and correlations, buying and selling positions in the various markets would be recommended. Unfortunately, one of the major criticisms of the Markowitz portfolio process is the unlimited degree of short-selling and investment in the risk-free asset. In many equity markets, restrictions on short-selling exist; derivative products help complete the market by providing an outlet for going short (which is as easy to do as going long). With futures, one simply places a selling order.

In options markets, the investor can create an equivalent short position by either purchasing a put option or selling a call option.

Another benefit of the Markowitz minimum-variance approach to asset allocation is that each investor can achieve the mix that is most appropriate to his own investment objectives or currency base. Thus, a German portfolio manager whose performance will be evaluated relative to a German equity index can construct an international portfolio that will provide the highest probability of superior performance relative to that index. In addition, the investor can also benefit from the fact that some derivative product markets can be mispriced. For example, in Finland, short-selling of the physical stock market is prohibited, resulting in the futures on the FOX (Finnish Option Exchange) stock index often being mispriced. In the recent past, FOX futures and call options were at substantial discounts, and simply purchasing the futures or call options (as a substitute to buying the FOX index) gave the investor a substantial increase in expected return (annualized of up to 25 percent relative to holding the stocks in the index). Therefore, by including these anomalies in the optimal portfolio selection process, those countries that present the greatest expected return will be allocated more investment. In addition, the minimum variance mix would not only select such opportunities but also would do so for the lowest level of expected risk.

Potential Trading Strategies Using Stock-Index Derivatives

In this final section, we examine two simple strategies using futures and options on international stock-market indices.

Consider a U.S.-based investor wishing to invest 10 million U.S. dollars in an international portfolio of equities, using these derivative markets. The investor would purchase the discounted equivalent of $10 million of three-month U.S. Treasury bills. If the prevailing rate was 9 percent, he would place $9,775,000 in the bills and would receive $10 million three months later. The investor would then split the remaining $225,000 into 14 different currencies and either post it as futures margin or purchase option contracts. In this example, we will assume that the investor will attempt to replicate the MSCIP World Index using the allocation mix suggested in Table 5. Therefore, he would leave 32.3 percent in U.S. dollars (or $72,900, which would be allocated to S&P stock-index futures or options) and convert 41.4 percent into Japanese yen (or $93,375, which would be allocated to Nikkei-Dow stock-index futures or options). This procedure would then be repeated for each of the other 12 currencies, and appropriate amounts would be invested.

Suppose the investor wishes to replicate the World Index using futures. He would buy futures to mimic holding the equity in each country. The investor would use his foreign currency endowments as the initial margin for the pur-

Figure 5 Buying Futures Contracts: Profit/Loss Profile

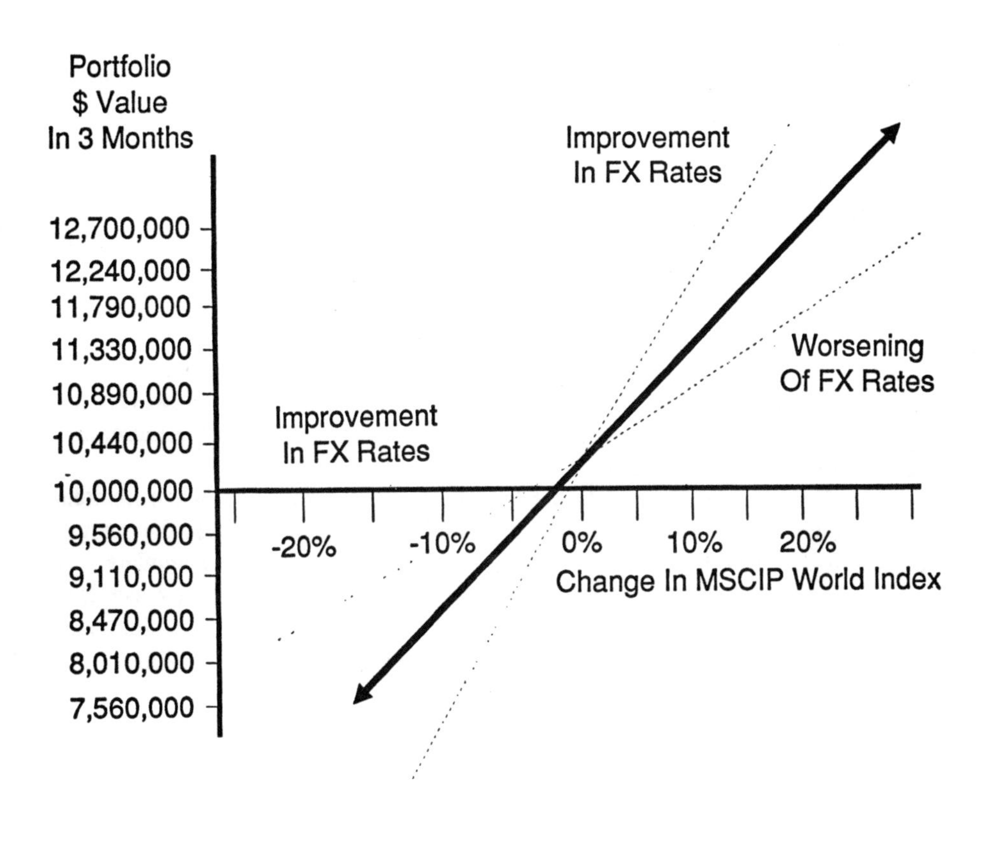

chase of these futures contracts. In our analysis, we assume the initial margin can be met with interest-bearing securities, and, therefore, the foreign currency endowments would earn some rate of interest (let us say 7 percent). Figure 5 displays the profit/loss profile for this strategy. If each stock market fails to move (that is a 0 percent change in the World Index) the investor will earn the 7 percent interest on the margin that would be returned. Of course, his final return will depend on the foreign exchange rate existing when the margin is repatriated. However, since 32.6 percent of the margin is in U.S. dollars anyway (32.3 percent in S&P futures and .3 percent in gold futures), the currency exposure of the initial margin is at most 67.4 percent. If the markets fail to move but the currency moves adversely, the final payoff for the investor will be made up of the

$10 million returned from the Treasury bill investment and the U.S. dollars returned from the unused margin accounts in the other currencies. While, he is exposed to movements in the currency markets, the exposure is limited to $151,650 in the margin accounts. This is a relatively small exposure, since if all the currencies move against him by 10 percent during the three-month period, that will represent a loss of only $15,165 in the margin value. If he had put all $10 million in actual equity, and the markets had been stagnant but the currency dropped by 10 percent, he would have lost $674,000 (the percent in the other markets multiplied by a 10 percent appreciation in the dollar). Of course, if the equity markets move either up or down, the futures positions will accrue profits or losses, and the currency exposure will be an increasing function of the amount that the world equity markets move. In Figure 5, that is why there is a dotted line diverging from the futures profit-and-loss diagram. This reflects the positive and negative effects of a 10 percent movement on exchange rates. When the markets are stagnant, the currency exposure is minimized. However, the cone shapes heading to the northeast and to the southwest reflect the increasing nature of the currency exposure.

A significant problem with the long futures program is that when markets move down, the investor must convert more U.S. dollars to meet the margin calls associated with his losses on these contracts. As losses accumulate, his capital will be reduced at the same time his currency exposure will increase. It is impossible for the investor to determine what the maximum loss potential could be on this strategy at initiation (apart from losing the entire $10 million). What of the options buying program?

Suppose instead the investor chooses to take his $225,000 and purchase three-month call options in amounts allocated to each market. Since all of his investment has been spent on option premiums, he will forgo any interest he might have received by posting the money as futures margin. However, he does know up front the maximum he can lose. Figure 6 displays the profit/loss profile of the call-purchase program. If the equity markets remain stagnant or fall, the call options will expire worthless. His currency exposure will be nonexistent if this occurs, since $10 million in U.S. Treasury bills will be returned, and the value of all his options will be zero. If, however, the equity markets rise, the options will finish in the money, and the investor can realize profits either via cash settlement at expiration or by selling the options on the last trading day. When any of the options have value, the proceeds must be converted back into U.S. dollars. As with the futures strategy, the greater the world stock-market increase, the greater the currency exposure. However, this exposure exists only for profits and not for the principal amount, which was left in U.S. dollars. In Figure 6, combination of stock-market profits and the associated currency exposure is represented by the cone-shaped pattern expounding to the upper right.

Figure 6 Call Option Buying Profit/Loss Profile

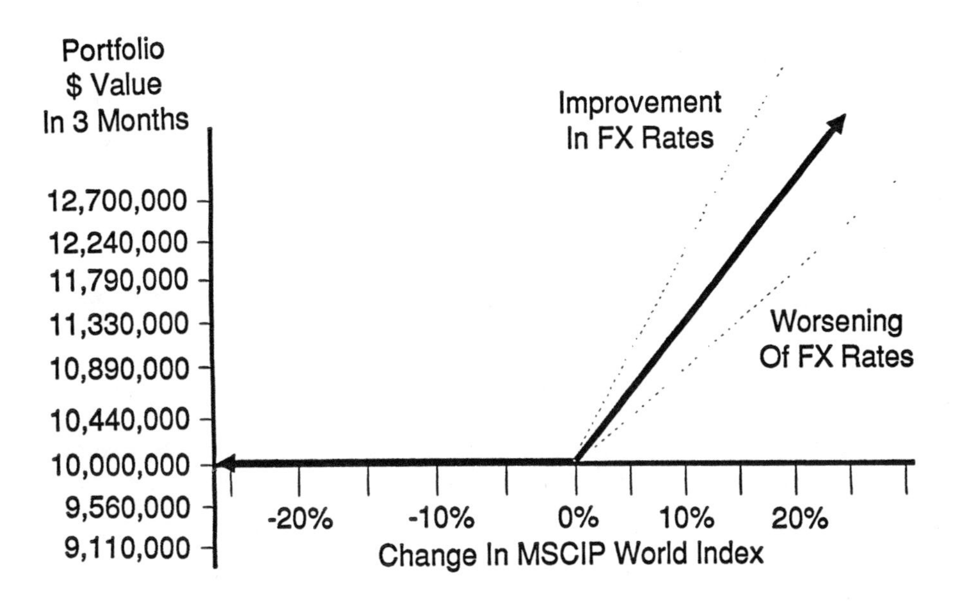

In these examples, we attempted to replicate the World Index. The investor could have just as easily sold futures or bought put options if his overall view of the world equity markets was pessimistic. In addition, he could have purchased both put and call options to benefit from increased volatility in world equity markets and would have achieved the straddle options strategy, well known in options trading circles. In fact, the investor has an unlimited number of possible trading strategies he can create by using both call and put options on stock indices around the world.

In this chapter, we have examined how the introduction of futures and options on stock indices around the world will assist the portfolio manager to implement an international diversification. As more and more countries add stock-index futures and options to their capital markets, international portfolio managers will have an increasingly wide range of tools to refine their approach to international diversification.

References

Aliber, Robert Z. "The Interest Rate Parity Theorem: A Reinterpretation," *Journal of Political Economy* 81, December 1973.

Black, F. "Capital Market Equilibrium with Restricted Borrowing," *Journal of Business*, July 1972.

Cornell, B., and K. French, "The Pricing of Stock Index Futures," *Journal of Futures Markets*, Spring 1983.

Fama, E.F. *Foundations of Finance*, New York: Basic Books, Inc., 1976.

Lessard, D.R. "World, Country and Industry Relationships in Equity Returns: Implications for Risk Reduction through International Diversification," *The Financial Analysts' Journal*, January-February 1976.

Lessard, D.R. "International Diversification." In *Financial Analysts Handbook*, 2d edition, edited by N. Sumner. Dow Jones-Irwin, 1988.

Stulz, R.M. "Pricing Capital Assets in an International Setting: An Introduction." In *International Financial Management*, edited by D. Lessard. New York: John Wiley & Sons, 1985.

CHAPTER 5

Combining Active Management with Indexing*

Seth M. Lynn
Axe Core Investors, Inc.

For those plan sponsors fortunate (or lucky) enough to have anticipated the opportunities, diversification of their pension assets into foreign stocks has provided outstanding results.

Without exception, international stock performance indices have shown an annual average return for the past three calendar years in excess of 40 percent. By comparison the Standard & Poor's Composite Index (S&P 500) logged in at 18.3 percent for this period.

As in the case of domestic equity managers investing in U.S. stocks, some international managers have bettered a representative index, but most have experienced a relative shortfall.

This phenomenon exists despite the general contention that international markets are less efficient, thereby affording active management with better odds for outperforming the averages. Such has not been the case, and one major factor in this failure is a function of foreign currency fluctuations.

* This chapter is an adaptation of Seth M. Lynn, "Combining Active Management with Indexing," *Investment Management Review*, September/October 1987. Copyright 1987 by DMA Communications.

1986 Returns

Introduced into the already enormous challenges of foreign security selection and market analysis is the variable of exchange rates, driven by market forces and manipulated by politicians. Fundamental security analysis helps managers evaluate the relative merits of alternative investments, but changes in the exchange rates can dramatically modify nominal returns—that is, those achieved from the perspective of local currencies.

For example, compare 1986 return for each of the foreign markets shown in Table 1. You can see that the nominal rankings for 1986 change when the numbers are converted to dollars. Spain, the best performing market in dollar terms, had its returns enhanced by nearly 40 percent. On the other hand, much of the 300 percent rally in Mexican stocks got lost in translation to dollars. Eleventh-ranked Belgium got a 90 percent boost from the dollar's decline. The most dramatic example is Switzerland, where mediocre market results jumped 700 percent as a function of currency moves. As any observer can see, the currency pendulum can swing either way.

Table 1 Major World Equity Markets 1986 Performance

Country	Local Currency (%)	Rank	Convert to $ (%)	Rank
Australia	46.75	8	43.10	11
Belgium	39.14	11	74.27	7
Canada	5.71	16	7.01	18
Denmark	–18.86	20	–1.55	19
France	49.66	6	76.34	6
Hong Kong	46.55	9	46.88	10
Italy	58.14	4	96.94	4
Japan	49.23	7	89.04	5
Malaysia	22.51	12	13.97	17
Mexico	312.32	1	102.61	3
Netherlands	8.92	15	38.04	12
New Zealand	99.23	3	111.99	2
Norway	–8.99	19	–6.68	20
Singapore	55.25	5	50.63	9
Spain	108.31	2	143.29	1
Sweden	41.52	10	58.30	8
Switzerland	3.83	18	32.96	14
United Kingdom	18.25	13	21.30	15
United States	14.62	14	14.62	16
West Germany	4.86	17	33.40	13

Our suggestion is that exchange-rate anticipation is a challenge faced by all international managers, and it is a task at which most would prefer to be lucky than smart. In terms of relative difficulty, conventional market timing is mere child's play. But as Table 1 demonstrates, there is considerable incentive to capture currency-leveraged performance.

For this reason alone, many capable international managers achieve only mediocre results. Their stock selection activities are finely disciplined and supported by an outstanding research effort. In a sense, they are able to walk on water with one foot.

Unfortunately, that is not enough in this competitive arena, and many without the necessary capabilities feel compelled to make these exceedingly difficult decisions. In effect, the manager is forced into a position that may be (and often is) detrimental to his client's portfolio.

Because of the dramatic volatility of currency fluctuations, the performance gains achieved through hard work are diluted or, worse, are entirely offset by fickle events beyond the scope of fundamental analysis. By trying to walk on water with both feet, managers often doom their clients' assets to results that are no better than those achieved by an inexpensive index fund.

The Solution

There is a simple solution to this challenge, but it takes cooperation among managers and their clients. It is based on a long-standing investment strategy used in domestic equity funds known as a *core-satellite* design.

In essence, an index fund—the core—is introduced into the plan to afford cheap, efficient diversification. Specialty managers then concentrate on their unique capabilities to achieve superior performance without concern for how their higher risk profile will affect the portfolio. The goal of the total fund is to harvest the performance advantage without incurring unconscionable risk.

The success of this strategy has encouraged some to consider its use for international diversification. The effect of currency fluctuations on portfolios' results requires a modification that offsets the risk and actually increases the potential benefits of this design.

Index funds are broadly available that replicate accepted foreign stock universes. Their results, within reasonable tolerance bands, can tract these indices. There is no shortage of capable active international managers, but more often than not, their investment results have recently been coming up short vis-a-vis market indices.

With a conventional core-satellite approach, results can be expected to be no better than those achieved by the index fund itself, and the fees would be higher. A slight modification in design offers plan sponsors an opportunity to achieve superior results despite the fact that alone, the active component of the fund achieves substandard results. All it takes is a coordinated effort and a non-parochial perspective.

Making it Work

Consider the $20-million international investment portfolio in Table 2. The active manager, using his skills at stock selection, purchases the shares he considers to be the most attractive for investment, regardless of the currency profile that results. Basically, he and the client must agree to ignore the effects currency translations have on the results of this $20-million portfolio. In essence they will view the fund's investment in local currencies and *not* translated into dollars.

Once the active portfolio is constructed, the core manager will neutralize all the peaks and troughs in the portfolio's currency profile using a complementary fund. In currencies where the active manager has substantial commitments, the complement will have limited exposure. Because of its mandate, the active fund undoubtedly will have gaps in its currency diversification. The complemen-

Table 2 International Portfolio Comparison

Country	Active Manager (-000-) $ mv	% mv	Market $ mv	Difference % mv
Australia	1,830	9.15	2.41	6.74
Austria			0.05	–0.05
Belgium			0.63	–0.63
Denmark			0.23	–0.23
Great Britian	3,174	15.87	14.99	0.88
France	772	3.86	3.43	0.43
Germany	3,488	17.44	4.79	12.65
Hong Kong			1.33	–1.33
Italy	644	3.22	2.64	0.58
Japan	7,568	37.84	61.77	–23.93
Netherlands	908	4.54	2.67	1.87
Norway			0.12	–0.12
Singapore	862	4.31	0.94	3.37
Spain			0.94	–0.94
Sweden	404	2.02	0.78	1.24
Switzerland	350	1.75	2.28	–0.53
Other Unlisted			0.00	
Total	20,000	100.00	100.00	.00

tary portfolio manager will, with the perspective of the total $100-million fund, invest a market commitment in a broad array of otherwise absent currencies.

A simple active-complement configuration could look like Table 3. The active manager has major over-weighted positions in Australia, Germany and Singapore. Germany was one of the poorer performing markets from a nominal perspective, but the other two were among the best performers as shown in Table 1. However, when the performance of each was converted to dollars, look what happened to Germany's results. On the other hand, investments in Australia and Singapore experienced a dilution in returns when converted. The manager invested in excellent markets, but currency effects hurt his results.

Furthermore, he entirely missed the golden opportunities in Spain, Belgium and Hong Kong and participated, to a limited extent, in the Yen's explosive rally.

Summing Up

Is this simply a hypothetical construct for the purposes of comparison? No. The portfolio actually is being managed by an active international manager. Its performance has, for the past few years, roughly approximated that of an international index fund. Unfortunately, the nominal returns, relative to the market, were excellent, but the currency profile had a detrimental impact on results.

Had a conventional core-satellite approach been used, the results would have been no better than those achieved by the market. The active component would have provided no incremental returns. On the other hand, had an active-complement strategy been implemented, the total fund's results would have improved. The stock selection capabilities of the active manager would have been segregated by the currency-effect neutralization of the complement, thereby promoting better total fund performance.

Why has this approach not been used by plan sponsors and/or active managers? For one, international investing is just getting a foothold in domestic pension funds. Competition is fierce among managers, and sponsors are still learning about the inherent risks and rewards of this alternative.

Furthermore, while currency effects can dilute good performance, they also can bolster otherwise mediocre results. Isolating a manager's capabilities could be considered a restriction of his flexibility. Saying one has an excellent stock-selection discipline is one thing; being held to it over time is another.

The approach of using an active-complementary structure will highlight the capabilities of an active manager. If he has the courage to foreswear currency timing activities and concentrate on security selection, this combined approach will benefit the total fund.

International managers may not be able to walk on water with both feet, but an active-complement configuration will help them find stones beneath the water's surface on which both feet can rest.

Table 3 Hypothetical Construct (Portfolio -000-) Active$20,000, Complement $80,000

Country	Active Manager (-000-) $ mv	% mv	Complement Portfolio (-000-) $ mv	% mv	Total Portfolio (-000-) $ mv	% mv
Australia	1,830	9.15	580	0.73	2,410	2.41
Austria	0	0.00	50	0.06	50	0.05
Belgium	0	0.00	630	0.79	630	0.63
Denmark	0	0.00	230	0.29	230	0.23
Great Britain	3,174	15.87	11,816	14.77	14,990	14.99
France	772	3.86	2,658	3.32	3,430	3.43
Germany	3,488	17.44	1,302	1.63	4,790	4.79
Hong Kong	0	0.00	1,330	1.66	1,330	1.33
Italy	644	3.22	1,996	2.50	2,640	2.64
Japan	7,568	37.84	54,202	67.75	61,770	61.77
Netherlandfs	908	4.54	1,762	2.20	2,670	2.67
Norway	0	0.00	120	0.15	120	0.12
Singapore	862	4.31	78	0.10	940	0.94
Spain	0	0.00	940	1.18	940	0.94
Sweden	404	2.02	376	0.47	780	0.78
Switzerland	350	1.75	1,930	2.41	2,280	2.28
Other Unlisted	0	0.00	0	0.00	0	0.00
Total	20,000	100.00	80,000	100.00	100,000	100.00

CHAPTER 6

Program Trading

Thomas Clark
Morgan Stanley

Michael Gibson
Morgan Stanley

Introduction

Recent reading of many current financial publications can lead investors to believe that there is only one force driving the markets nowadays: program trading. It appears to affect stocks in both the United States and abroad like no other current phenomenon, pushing prices in all directions with no obvious rationale. It has been singled out by a few large investors as detrimental to the marketplace, and pleas have been made to the U.S. regulators to stop the practice.

As is often the case, the first glance is not always the correct one. The terminology involved is often as misunderstood, as are the various components that use program or portfolio trading. The purpose of this chapter is to shed a little light on some of the realities behind the myth.

Program trading was the result of an evolutionary process responding to the changing needs of investors in the securities markets. After World War II, the main characteristic of the equity markets was individuals trading small amounts of stocks for their own account. In the 1950s the rise of pension plans and their 30 percent exposure to equities, as well as the development of mutual funds, led to markets' needing transactions involving larger numbers of shares. Executing large orders (100,000 + shares) would take weeks. It was this lack of liquidity that brought about block trading.

To facilitate transactions, large brokerage firms had to offer customers immediate execution of sizable blocks of shares, even if this meant taking the opposite side on some transactions and keeping this position on their books until they could satisfactorily dispose of it in the market. This meant also that the broker became a dealer by putting his capital at risk. But block trading also brought additional liquidity to the markets, which in turn attracted institutional players looking for rapid execution. The development of block trades was rapid, growing from 3 percent of the market volume in 1965 to 50 percent in 1986.

Parallel to the increase in the average transaction size, institutional customers started to look for swapping capacity, directly exchanging one stock for another. The customer would get out of GM and into Ford in one operation, thereby retaining a similar exposure to a given industry. This then evolved into the swapping of groups of stocks to change the exposure of a portfolio to one or many sectors of the market.

In the mid-1970s, the growing popularity of random walk, modern portfolio theory and disappointment with the performance of active managers led to the development of index funds based on the S&P 500. This provided demand for portfolio-level transactions and other innovations. Furthermore, the advent of negotiated commissions in 1975 meant that customers could get brokers to bid competitively on portfolio trades, reducing the overall cost to the customer.

As institutions (essentially pension and mutual funds) became larger players in the marketplace, they required the ability of brokers to execute increasingly larger transactions. The markets responded with mechanisms that catered to the new demand.

In the 1980s, institutions discovered the advantages of asset allocation, switching between stocks, bonds, and cash to take advantage of higher forecasted returns. This strategy required brokers who could complete complex transactions rapidly and efficiently. This was a step further in the evolution from individual share trading toward portfolio-type transactions.

The NYSE has now introduced the Exchange Stock Portfolio (ESP), a one-stop trade in which customers can purchase or sell a portfolio of stocks that exactly represents the S&P 500. Several well-capitalized Wall Street firms act as market makers (the $5 million required per trade being too large for the existing stock specialists), with transactions being quoted over the NYSE ticker.

Much of the criticism of program trading stems from the accusation of the increase in volatility that it induces in the market. This is not the case, as can be observed in market data. The average market volatility (as expressed by the NYSE composite index) for the first 10 months of 1989 (0.113) is lower than the average volatility of the last 15 years (0.1437). In fact, five out of the last six years have been characterized by below-average volatility.

Additionally, the size of a block trade (10,000 shares) in relation to the average daily NYSE volume during the 1960s (6 million shares/day) is about the same as the relation of an ordinary portfolio trade (150,000 shares) to the average daily NYSE volume during the 1980s (97 million shares/day). In other terms, portfolio trades are to the 80s what block trades were to the 60s, with both taking a 15 percent of the market volume after a few years of existence.

Many investors ignore the fact that an 80 point move with the Dow at 2700 is not the same as an 80 point fluctuation with the Dow at 1000. Many investors refuse to recognize that derivatives, in acting as hedging instruments, have vastly enhanced the liquidity of the underlying stock market, which in turn allows the same managers to move ever-increasing amounts of capital they invest in the market. They also ignore that many of the more sophisticated money managers use one or many of the portfolio trading components to their advantage. They execute their trades rapidly and in a cost-effective manner through portfolio trades and use index arbitrage to enhance the return on their cash positions.

Let us return to the development of the market for portfolio trades. The introduction of various derivative products (futures and listed option contracts) led to more efficient pricing and hedging capabilities, especially for customers dealing at the portfolio level. Listed options on individual stocks were introduced in April of 1973. Futures and options on indices appeared one decade later: Value Line and S&P 500 in 1982, S&P 100 in 1983. These instruments brought with them many advantages. They increased the liquidity of the stock market, simplified the hedging process and lowered commissions.

The following table shows the advantages in using futures over the underlying stocks in the world's main securities markets. This shows, for example, that in Japan the market impact (price reaction to the supply and demand for a given security, generally a function of liquidity) is 12 times smaller when transacting in the futures rather than the stock market.

Ratio of Stock to Futures Cost

	United States	Japan	United Kingdom
Commissions	10:1	6:1	16:1
Market impact	9:1	12:1	26:1

As many portfolio managers rapidly observed, there were and still are, significant economies associated with the use of derivatives. The next table compares the average daily volume for the same markets.

Daily Volume in Billions of Dollars

	United States	Japan	United Kingdom
Stock market	$ 6.6 (NYSE)	$ 11.2	$ 2.4
Futures	$ 12.2	$ 4.1	$ 0.2

The derivatives have, therefore, become an integral part of world markets. They cover 93 percent of the stocks figuring in the Morgan Stanley Capital Investments (MSCI) World Index, allowing global investors to hedge large portions of their worldwide portfolio.

Numbers published by the NYSE since July 1988 (based on reports of member firms) show that program trading generally averages between 10 and 11 percent of total NYSE trading volume. The figures published by the NYSE are somewhat misleading in that they present the amount of shares involved in program trades over the total volume of transactions. However, while for every purchase there is a sale when computing total volume, the same cannot be said for program trades, because they generally take only one side of the transaction. So the figures put forth by the NYSE should in fact be divided by two to obtain the correct market share of portfolio trading.

Index arbitrage, which will be described below, represents one-half of the total amount, with the other half split between other strategies. NYSE members also indicated that they used program trading on other domestic exchanges as well as on foreign exchanges, albeit to a lesser extent.

Foreign Countries Now Offering Index Futures Contracts

Japan	Nikkei 225*, TOPIX*, Osaka 50
United Kingdom	FT-SE 100*
Singapore	Nikkei 225* Hong Kong: Hang Seng
Australia	Australia All-Ordinary*
Netherlands	EOE*
France	CAC 40*, OMF 50
New Zealand	Barclays 50
Canada	Toronto 35, TSE 300

* Index options available.

At the time of writing, some countries such as Switzerland and Sweden offer only index options but no index futures. However, the list above is constantly being enlarged with new countries adding derivatives to their financial markets.

What is Program Trading?

Seldom has financial terminology led to such confusion and misinterpretation. In essence, program trading (or portfolio trading, which better reflects the activity) is simply the mechanism through which different investment strategies or techniques such as index funds, index arbitrage, asset allocation, and portfolio insurance (all of which will be further discussed) are executed.

The term *program trading* originally referred to the list of hundreds of stocks that index funds wanted to buy or sell that day. The list was a computer printout of names dictated over the phone. The New York Stock Exchange itself identifies a program trade as any simultaneous transaction in more than 15 stocks.

One basic difference between stock and portfolio trading is that decisions governing the latter are generally based on macroeconomic factors and not the performance of individual stocks. Portfolio traders do not react to disappointing earnings of one company.

Many people have associated the use of program trading with automated, computer-driven techniques that obviate the intervention of human beings. In other words, the computer makes all the decisions. We shall see that this is not the case. However, computers are used to simplify the mechanics of the process. They digest large amounts of data and produce synthesized real-time reports. They allow rapid execution through automated trading systems such as the NYSE's SuperDot. This system forwards trade instructions through a computer link to the specialist on the floor of the exchange. The specialist then has to manually confirm the trade. This system currently handles 70 percent of the NYSE trades.

The GLOBEX network, soon to be introduced by the Chicago Mercantile Exchange and Reuters, is a glimpse at the markets of the future. This off-hours trading system will soon lead to 24 hour trading of securities and commodities through one centralized vehicle.

We'll now review the various techniques associated with program trading.

Index Arbitrage

The purpose of this strategy is to take advantage of discrepancies appearing between the price of an index futures contract and the underlying stocks that make up the index. The difference in price between the two is usually adjusted for both the dividend on the stocks (not available to the holder of the futures contract) and the interest rate at which one can invest the amount saved through buying futures on margin rather than the underlying stocks. These adjustments, determine the "fair value" of a futures contract. If a contract is not priced "fairly," one buys the cheap side and sells the expensive one, knowing that upon expiration both sides will be of equal value, regardless of the level of the stock index.

The existence of an index futures contract is not a prerequisite to conduct index arbitrage. If European options exist on the index, synthetic future contracts can be created through a combination of puts and calls. However, synthetic futures are not as flexible as normal futures, because option exercise prices do not provide such comprehensive price coverage as the individual ticks of futures contracts. The S&P 500 futures contract moves by increments of 0.05 index points, while the options exercise price jumps every 5 index points.

Stock-Index Arbitrage

Two strategies have emanated from index arbitrage. The first is stock-index arbitrage. The purpose of stock-index arbitrage is to generate returns above the level available on U.S. Treasury bills. This is why many participants call index arbitrage an "interest rate" operation. If futures sell above their fair value, the arbitrageur (who initially owns a cash position) buys the stocks in the index and sells futures. Recall that we adjusted the index for dividends and the interest rate to obtain the fair value of the futures. Assuming a known dividend, we can translate a futures contract with a price above its fair value into an above-normal interest rate. The arbitrageur is hedged, because he knows that both investments will be of equal value upon expiration. If the price of the future approaches the price of the underlying stock before expiration, the strategy can be unwound early, yielding even higher returns.

Index-Fund Arbitrage

The second strategy is index-fund arbitrage. The purpose of this strategy (the reverse of stock-index arbitrage) is to boost returns of institutions that own "index" funds by exploiting discrepancies between the futures and the underlying stocks. If the price of the futures drops below the value of the underlying stocks (which determines the fair-value price of the futures contract), the index fund sells part or all of its shares and replaces them with futures (retaining at all times the same equity exposure). The arbitrageur has just replaced his physical portfolio with an identical synthetic portfolio at a lower cost. When the futures come back in line, the strategy is unwound by selling futures and buying the stock.

Both types of index arbitrage can be done either for clients or for a broker's proprietary account. Numbers published by the NYSE show that most activity is for customer's accounts, refuting the argument that this is essentially a game played between people on Wall Street. In October of 1989, most firms withdrew from proprietary index arbitrage in the United States.

The economic justification of index arbitrage is that it allows people who hedge using futures to be reasonably certain that the derivative will fluctuate in line with the underlying stock. This gives people confidence in the hedging instrument, which leads to more equity investors and higher liquidity.

Over time, the inefficiencies of the market have decreased to a point where it is increasingly difficult to take advantage of discrepancies. In the United States, S&P 500 arbitrage spreads that used to be 0.6 of an index point two years ago are now down to 0.2. Opportunities must be looked for abroad.

The concept of index arbitrage developed in the United States when futures were created in 1982. However, it is not confined to North American markets. It has extended to Japanese markets, where the first expiration of the Nikkei futures contract caused some emotion among the locals. The Japanese index arbitrage market is dominated by dealers with only a few customers participating. The presence of U.S. firms, as well as cross-border joint ventures and participations (such as Nikko Securities stake in Wells Fargo Investment Advisors) ensures continued development of portfolio trading activities in Japan.

In theory, index arbitrage is possible on any exchange with an index futures or options contract. However, one must account for the particularities of executing equity and derivative trades in each country.

In the United Kingdom, high costs (nondealers are subject to a turnover tax of 50 basis points), added to the fact that certain clients cannot borrow securities (to go short), mean that this activity is left to dealers. In the Netherlands, an efficiently priced futures market makes index arbitrage difficult to accomplish despite good liquidity. France has seen occasional outbursts of activity in a dealer-driven market, generally acting with baskets rather than entire indices. High transaction costs in countries such as Switzerland (low liquidity, quasi exclusively dealer-driven market and very high charges) and Sweden (2 percent tax each way and prohibition of shorting) render arbitrage activity uninteresting. Hints of index arbitrage appeared in Hong Kong, but activity seems to have decreased since the 1987 Crash. Costs of operating on the Australian markets often prohibit such activities.

Index arbitrage has been subject to a lot of scrutiny, especially after the Crash of 1987. The NYSE prohibited the use of the automated trading system once the Dow Jones Industrial Average moved more than 50 points in one day. Manual execution of orders took over, giving an advantage to the larger firms with a well-organized floor execution capability. The rule was discontinued six months later without having really been put to test. The various enquiries that followed the crash exonerated index arbitrage as being a culprit for the market drop. After the dip of October 1989, other circuit breaking measures were implemented. Their efficiency remains to be tested.

Index arbitrage is a linking mechanism that strengthens the correlation between an index and its derivatives. A strong correlation allows hedgers to execute many of the following strategies.

Portfolio Insurance

This technique, also called dynamic asset allocation, is used by institutional investors who wish to protect their portfolios against market declines. Its theoretical foundation comes from the theory of options. It seeks basically to replicate the structure of a put option, switching funds gradually into equity as the market rises and into cash when the market declines. The sale of stocks can be replaced by the purchase or sale of futures contracts.

Portfolio insurance requires selling into weakness and buying into strength, thereby amplifying the movement and the volatility of the markets. This sends the market both higher and lower than it would have been otherwise.

Portfolio insurance did not stand up well to the test of the crash, because the confusion created by the lack of timely information led to the withdrawal of index arbitrageurs, which in turn created very large discrepancies between the futures and the underlying index. In addition, the Black & Scholes formula, upon which the portfolio insurance is based, presupposes orderly markets that allow timely trading, as well as constant volatility. This was obviously not the case on October 19th.

In the United States, portfolio insurance is now estimated to cover approximately $10 billion, down from above $60 billion before the crash. Other instruments, such as long-term index puts, are starting to appear, filling a gap in the market that led to the development of this strategy in the first place. However, some of the put writers are using dynamic hedging to cover themselves.

Portfolio insurance, despite its inherent weaknesses, is starting to appear sporadically in Japan and the United Kingdom.

Asset Allocation

Becoming more and more popular, this form of investing, also called tactical asset allocation, is similar to what used to be called market timing. One difference may be the use of computers and analytical models in attempting to predict returns. Managers who take a top-down approach to investing often wish to move rapidly from one asset category to another. They want to do this in one transaction, taking advantage, for example, of the higher return that they forecast in the equity market as compared to the bond or cash markets. This can be done either directly in the market or through the use of futures, taking advantage of their larger liquidity. Speed of execution is generally important to this strategy. Depending on the model, shifts can be sizable, moving (in extreme cases) from 100 percent equity to 100 percent bonds or cash in one trade.

Index-Fund Management

Regardless of one's persuasion regarding "top-down" or "bottom-up" methods of investing, the sheer size of some portfolios renders the traditional approach to stock picking inappropriate. An outstanding performance of one stock in which a manager holds a $5 million stake will do little to affect the overall return of a $150 million portfolio. If one considers the additional effort, analytical resources, risks, and costs involved, it makes sense to consider indexing for at least one portion of the overall portfolio.

Managers choosing indexing will generally use portfolio trading methods to enter the index, as well as to cope with further sizable cash infusions or withdrawals.

Indexes are not selected because of their ease of trading but for their representation of a market or a subcategory of the overall market. Investing in all the stocks of an index often does not make sense because of the number of stocks involved, the illiquidity of certain stocks, or even outright restrictions, especially in foreign countries. The solution to these problems has been found in the form of baskets of stocks. These baskets are developed using analytical techniques. They are easy to trade and track the performance of their underlying index closely.

Manager Transitions

In transiting from one manager to another it is important to keep transaction costs as low as possible. These costs can be minimized through the use of an index-arbitrage trade. The old portfolio is sold when stocks are expensive relative to the futures that are purchased instead. The new portfolio is bought when stocks become cheap relative to the futures. This reduces transaction costs and maintains market exposure at all times.

The portfolio trade will also allow you to pinpoint the performance of a manager to a specific date rather than have a transition drag on over a period of time. First, perform an analysis of the old portfolio for stocks to sell, determine which shares to buy in replacement, then establish a buy-and-sell list for brokers to bid on. The bidding process will be described later in this chapter.

Additionally, when managers receive funds but are not ready to invest, they can use futures to maintain market exposure until they are ready to purchase selected stocks. The exposure will enhance their ability to track the benchmark against which their performance is measured.

Corporations (plan sponsors) may also want to terminate a pension plan that is overfunded to put the money to better use. A portfolio trade will expedite the trade execution.

Quantitative Rebalancings

An alternative to investing in an index is to purchase a basket of stocks that replicate the index. This is also known as core-indexing. The stocks are chosen for the way in which they, as a group, track the index as closely as possible. Investing in a basket generally leads to selecting liquid stocks, which in turn decreases transaction costs. Core-indexing is useful to smaller investors who may not be able to acquire all the stocks of an index, especially when investing in foreign stocks.

Core portfolios are rebalanced when monitoring indicates significant changes resulting from external factors (change of the components of the underlying index) or mistracking (the stocks selected for the basket cease to closely track the index). A portfolio is rebalanced taking into account the trade-off between expected tracking accuracy and transaction costs. The trade that ensues sells the unrequired stocks and replaces them with those that ensure optimal tracking of the rebalanced portfolio.

Execution

Approach

For a broker to excel in worldwide portfolio trading, he must be able to assess the risk involved in the program he is asked to bid on. The portfolio must be assessed in terms of liquidity, volatility, and sector attribution. More specifically, special attention should be given to evaluating and understanding the trading of the less liquid names in the portfolio, also called the "tail." It is these stocks that will be the most complex to trade and will often lead inexperienced brokers to bid too high based on the perceived higher risk.

The broker must also be able to provide advice as to hedging techniques the customer may wish to use. For example, he can use derivatives to obtain or cancel market exposure, then proceed with the individual purchases or sales of stocks at his leisure. In many foreign countries, futures and options are still at an embryonic stage. Their bid-ask spreads are often significant, and the basis difference is often large, which would lead to higher execution costs. Professional advice becomes a fundamental part of the analysis.

The broker must also have a 24-hour multicurrency distribution capacity, including a presence in all major markets and a significant volume of business in the smaller ones that guarantees them a correct execution from the local broker. Weak execution in one market could well offset profits made in another.

Executing these complex trades requires the technology necessary to transact efficiently thousands of trades in 18-20 countries worldwide and report them within a 24-hour period. The broker's operational setup should be considered by the customer as a crucial element of the trade package.

To the investor, the advantages of portfolio trading are that he has a guarantee that the whole trade will be completed without a problem. Unless he chooses an agency execution, he will also enjoy a price determined in advance and minimum market impact resulting from the confidentiality of dealing through one broker. In all cases, he will be in contact with one broker who will synthesize the worldwide execution.

Packaging

There are many types of packages available for portfolio trades. A peculiarity is that they are often offered to different brokers who are asked to bid on them. The commission depends primarily on how much the customer wishes to reveal. The more revealed to the broker, the cheaper the commission. A main advantage to the portfolio trade is that the broker will guarantee the execution price in all cases except agency bidding.

Agency bidding is the least expensive way to execute an order. No guarantee is given as to the price. This is very similar to an ordinary stock trade with a volume discount, except that it involves portfolios rather than individual stocks. It is the solution that most customers prefer to use for large international portfolios. To illustrate the difference between the various methods, let us use an example involving the purchase of 100 stocks in three countries. Customer XYZ will put together the list of stocks and provide it to one or more brokers to bid on. They will quote a number, generally expressed in basis points, that will represent their commission above the price obtained.

Open-hand bidding guarantees the price for a package in exchange for slightly higher commissions. The price is based on an agreed point in time, such as the market close. The customer identifies the stocks and the number of shares to the broker before the bid takes place. The broker is generally given a time limit to complete the trade. To follow up with our example, customer XYZ gives the list of stocks and the day of execution to his/her broker/s, who will then bid for the execution. The price will generally be that of the next closing price of the individual markets worldwide plus/minus the brokers bid.

Blind (or closed-hand) bidding guarantees a price prior to the customer identifying the individual stocks or the number of shares on the list. General characteristics of the list are known. The moment at which the customer reveals the individual names is established in advance. The cost per share can vary considerably depending on the parameters identified. The more information provided, the cheaper the commission to the customer. Here, XYZ will not hand out the individual stock names but rather describe the list, indicating the percentage per country, per industry, the average beta, etc. They will also indicate the time at which they wish to execute and how much in advance the stock names will be made available. The broker/s will bid, and the winner will receive the list of the stock and time of execution.

Double-blind bidding has the highest commissions (high end of blind-bidding range). It is similar to blind bidding except that the time of delivery of the list of names is unknown to the broker. XYZ will again provide only general parameters on the stocks. However, this time they will no indicate when they want to execute the trade, only how much notice they expect to provide the winning broker.

Incentive trades have negotiated commissions that vary according to performance. They are structured to reward the broker for improving on the benchmark price of the portfolio. The price may be guaranteed, with a negotiated split for any improvement beyond the agreed price. Another alternative is to have the brokers commission rise with any improvement on the benchmark.

Globalization of Trading

There is sizable interaction between international investing and program trading. Many investors are interested in participating in the higher returns available in foreign markets but face considerable problems in selecting individual stocks. It is often difficult to obtain sufficient knowledge regarding the accounting and legal difficulties that may affect foreign stock prices. Other investors are interested in betting solely on the macroeconomic fundamentals of a country or group of countries and do not wish to be bothered with stock-selection problems (country active/stock passive managers). The solution to these problems is investing in a portfolio of foreign stocks that fully or partially replicates a given index.

Another reason for the growth in international investing is the recent development of worldwide indices that allow managers to compare their performance. These indices are structured to be comparable across countries. They also allow industry comparison on a cross-border basis. These indices, such as the Morgan Stanley Capital International "Europe Australia and Far East" (EAFE) Index are the foreign equivalent of the S&P 500 in the United States.

Investing in the index derivatives of foreign countries is also an excellent way to maintain equity exposure in the foreign stock market and only be subject to partial foreign exchange risk (the option premium or the futures margin).

Finally, as commissions and taxes decrease, so does the cost of entering or exiting the individual countries. This has led to an increase in the velocity of cross-border investing, facilitated by the process of portfolio trading.

Barriers

The rules of local exchanges often preclude markets from reacting in a similar manner. Countries that have a specialist system generally have fluid markets. Those with mechanical means of multiple execution in the marketplace further allow rapid, cheap, and efficient transacting. Other markets work in a manually written ticket environment that delays operations considerably and creates uncertainty as to the outcome of program trades.

Different markets have different closing rules that need to be understood before transacting. Collection and maintenance of dividend information is important when establishing the fair value of the future contracts. The overall cost of clearing and settling also needs to be known.

Conclusion

Program trading is the result of an evolutionary process put in motion by a need to provide the institutional market with significant liquidity. This increase in liquidity decreases the possibility of market manipulation.

The oncoming of futures and options has led to the narrowing of spreads as well as the reduction of transaction costs. Stock index arbitrage ensures that futures remain in line with the underlying stocks, leading to more efficient markets. Even if investors perceive the current environment as more volatile (which we have shown not to be the case), they should understand that, on average, the price level will be the same. By using limits on their orders, they can sell higher and buy lower than otherwise.

It is true that the stock market has been "commoditized" by the arrival of these new products. However, it was the disappointment over active managers' performance that led pension funds to turn to index funds and quantitative investment techniques. Nowadays, the sheer size of assets under management in many funds has made passive investment a necessity. This has led to the development of computer-driven, low-cost strategies. These strategies, often developed in the United States, have now spread throughout the world, aided by the development in global communications. It is too late to turn back the clock. These techniques are with us to stay. They will not, however, be the end of the individual stock picker, who will always play a role in evaluating the marginal price of the equity and arbitraging any discrepancy.

A lot of criticism comes from money managers who lack the understanding of these new instruments and see them as challenging their professional abilities to manage effectively. It is they who should take the time and the effort to fully comprehend the changing environment in which they will have to operate in the future, and then use these tools to their advantage.

In 1894, investors were prohibited from using the telegraph to invest in the stock market. In 1929, investors were forbidden to use the telephone to execute margin trades. Both these tools are now commonplace in the business world. Based on historical evidence, it seems that it will be only a matter of time before the negative feelings regarding portfolio trading become a phenomenon of the past.

In prohibiting large investors from transacting in size, they will be obliged to turn to other solutions to fulfil their needs, such as using foreign markets or crossing networks. This will in turn hurt the small investor, who will not have access to these alternatives and, therefore, have to transact in markets with lower liquidity and larger spreads.

The world of investment is changing, and performance is coming under increased scrutiny. For both markets and investors who do not wish to take the time and effort to understand the benefits of the techniques explained in this chapter, the horizon promises to be cloudy.

CHAPTER 7

The Performance of Currency-Hedged Foreign Equities

Lee R. Thomas III*
Investcorp Bank, E.C.

Risk and Diversification

This chapter examines the role currency hedging can play in managing internationally diversified equity portfolios. For the U.S.-dollar-based investor, diversification across foreign markets has proved to be a powerful way to market specific risk. But there are potential drawbacks. As you diversify into foreign markets, you assume exchange rate risks. And because dollar exchange rate changes are highly correlated, it is not easy to diversify away your currency exposures. Even if you spread your offshore equity holdings among many foreign markets, currency risks remain important. This chapter will help you evaluate both the riskiness of the implicit currency exposures that are embedded in an international equity portfolio and the kinds of strategies you can use to neutralize those exchange rate risks.[1]

* Lee R. Thomas III is Vice President of the Management Committee of Investcorp Bank, E.C. This article was written when the author was an Executive Director in the Financial Strategies Group of Goldman Sachs International, Ltd. The author wishes to thank Fisher Black for his comments, and Ron Knight for editorial assistance.

[1] Also see Lee R. Thomas III, "Currency Risks in International Equities," *Financial Analysts Journal*, March/April 1988, for a complementary analysis that evaluates currency hedging when equity risks are measured by beta. As in this report, the perspective is that of a dollar-based international investor.

Specifically, we will explore whether currency-hedged foreign equities are better diversification vehicles than unhedged foreign equities. Our research indicates that, while it is impossible to predict beforehand whether currency hedging will increase or decrease the long-run *return* on your portfolio, currency hedging has consistently reduced portfolio *volatility*.

In light of these results, we recommend that you use currency-hedged securities as your "base case" in formulating your international investment strategy. First, select international investments based on their intrinsic appeal, as if you intend to currency hedge. Then decide whether currency hedging is appropriate.[2] Bear in mind that not to hedge your currency risks amounts to adding a speculative foreign exchange position to your underlying foreign equity holdings. This you should do only if the return to currency speculation—based on your exchange rate expectations—is commensurate with the risks of bearing currency exposures.

We will begin with a review of the returns and risks of holding foreign equities. We look at the performance of six major foreign equity markets from 1975 through mid-1988, contrasting hedged and unhedged foreign stocks with U.S. equity investments. We also consider the gains a U.S. dollar investor would realize by diversifying into foreign markets, again comparing the hedged with the unhedged case. We then take a look at some of the practical problems associated with currency hedging an equity portfolio, and we evaluate the efficiency of hedges designed to solve those problems. In the final section, we summarize our results and discuss some of the implications of our findings for managing portfolios that include foreign equities.

Performance Report

Because our study extends from 1975 through June 1988, it includes most of the recent floating exchange rate period. The markets we cover—Japan, Germany, the United Kingdom, France, Switzerland, and Canada—account for about 88 percent of the current nondollar world market capitalization.[3]

Assumptions

We define unhedged returns in our study as the percentage changes of the dollar-measured equity-index values.[4] To compute the monthly hedged returns, we as-

[2] If you are investing in a foreign market *because* you want the currency exposure, our advice is: Don't. Instead, if you like a currency but not the underlying asset markets, simply add short-term foreign currency deposits to your portfolio. This will give you a pure currency play.

[3] Based on the market capitalization weights used to construct the FT-Actuaries Indices.

[4] Our equity returns data are based on the FT-Actuaries Indices and include dividends. FT-Actuaries Indices are market-capitalization-weighted, comprehensive indices of the tradable shares in each foreign market.

sume that at the beginning of each month, the investor sells forward (for U.S. dollars) for one-month delivery the foreign currency value of his equity shares. This locks in a dollar exchange rate on the initial value of his foreign stock holdings. But the hedge is imperfect. Even though the investor eliminates most of his currency risk, he remains currency exposed on the position's monthly profit or loss. We calculate exchange rate gain or loss on this unhedged exposure by comparing the beginning and ending spot exchange rates.

The hedge design we simulate is imperfect because the investor hedges only the *initial* value of his foreign holdings each month, rather than the *final* value. Moreover, the hedge is designed to eliminate the portfolio's *accounting* exposure, and this may not be the same as its *economic* exposure.[5] However, the strategy we simulate does conform to the hedging strategy many international investors use in practice. It is simple to implement and practical to administer.[6] Later on, we will investigate how seriously flawed our simulated hedging strategy actually is by comparing it to a theoretically perfect hedge. We will also consider whether using an alternative to our hedge ratio of 1.0—i.e., hedging exactly by the beginning-of-month face value of our foreign equity holdings—would improve our portfolio's performance.

Individual Foreign Market Performance

How have foreign equities performed from a dollar investor's perspective? Table 1 compares the returns and risks of the six foreign markets, a representative foreign stock portfolio, and the U.S. equity market over our sample period.

Considering returns only, some foreign equities have performed quite well. Three of the foreign stock markets we examined—Japan, the United Kingdom, and France—recorded higher unhedged returns than the U.S. market. In the three other cases, however, the foreign market's return was less. On average, the foreign markets returned 16.5 percent per year, bettering the average return of 14.6 percent for the U.S. market. Currency hedging increased the return in three of six cases (Germany, Switzerland, and Canada); substantially reduced the return in one case (Japan); and left return essentially unchanged in two cases (the United Kingdom and France). On average, the return on hedged foreign equities was virtually identical to the average unhedged return—16.4 percent versus 16.5

[5] To be precise, we do not take into account the correlation, if any, between stock prices and exchange-rate changes.

[6] It also has low transaction costs, since a single, standard foreign-exchange transaction, called a spot to one-month swap, acquires the foreign currency needed to buy foreign equities and simultaneously hedges them. Subsequently the investor rolls his currency hedge forward by executing a new foreign-exchange swap each time his old hedge expires.

[7] Here we regard volatility, measured as standard deviations of monthly returns, as a measure of risk, and historical volatility as a good proxy for prospective risk.

Table 1 Returns and Risks* (January 1975–June 1988)

	Unhedged			Hedged			
	Average Return (%)	Risk (%)	Return/Risk Ratio	Average Return (%)	Risk (%)	Return/Risk Ratio	Change in Risk (%)
U.S.	14.6	16.3	0.90				
Japan	23.9	20.4	1.17	20.7	14.8	1.40	–27
Germany	13.3	21.0	0.63	14.1	17.4	0.81	–17
U.K.	22.6	27.2	0.88	22.4	28.8	0.94	–12
France	16.0	26.3	0.61	15.8	22.2	0.71	–16
Switzerland	13.1	20.2	0.65	14.0	16.8	0.83	–17
Canada	10.1	21.3	0.47	11.4	19.4	0.59	– 9
Non-U.S. Average	16.5	22.7	0.78	16.4	19.1	0.88	–16
		17.0					
Foreign Portfolio**	17.7		1.04	17.3	18.8	1.26	–19

* "Average Return" is the continuously compounded annual rate of return. "Risk" is the annualized standard deviation of monthyl returns.

** The weights for our foreign stock portfolio are as follows: Japan, 30%; Germany and the U.K., 20% each; France, Switzerland, and Canada, 10% each.

percent, respectively—and almost 2 percent per year better than the return recorded by U.S. stocks.

But while foreign markets averaged a higher rate of return than the U.S. stock market, they were also more volatile. Held unhedged, each of the foreign market indices we examined suffered from significantly greater volatility than did the U.S. market. The annualized standard deviation of returns for the foreign stock markets ranged from 20.2 percent for Switzerland to 27.2 percent for the United Kingdom, averaging 22.7 percent for all six. This compared with a 16.3 percent annualized standard deviation for the U.S. market.

As we shall see, however, foreign stocks' exchange rate risk accounted for a good part of this difference. You can eliminate exchange rate risk through hedging. And you can reduce the volatility of foreign returns even further by diversifying your investments among the different offshore markets.

The Impact of Currency Hedging

Currency hedging was successful in reducing volatility in every one of the foreign markets we examined, and often the reductions were considerable. The most significant decrease in volatility we noted was in the case of Japan, where the annualized standard deviation dropped to 14.8 percent from 20.4 percent—a 27 percent reduction. Overall, hedging reduced the average volatility for the six countries in our sample by 16 percent, down to 19.1 percent compared with 22.7 percent unhedged. But even with currency hedging, five of the six foreign stock markets remained more volatile than the U.S. Market.

We get a slightly different perspective when we look at risk and return together. For each market we computed a simple measure of overall performance by dividing average return by the corresponding standard deviation of return. Now we can see that unhedged, all but one of the foreign markets recorded performance ratios lower than that registered by the U.S. market. The average ratio for the overseas markets was 0.73 for the foreign markets versus 0.90 for the United States. Hedging raised the return-to-risk ratio for every foreign market (although the ratios for four of the six still remained below that of the U.S. market). And the average of these foreign return-to-risk ratios improved by about 21 percent to 0.88. So, unhedged, the average foreign market substantially underperformed the U.S. market, measured by return-to-risk; hedged, the average foreign market's performance was about the same as that of the U.S. market.

Foreign Equity Portfolios

Most dollar investors who own foreign equities hold portfolios of those stocks rather than single issues. This is a wise strategy, because the risks of the various foreign markets tend to offset each other, reducing the risk of the portfolio as a whole. So instead of focusing on the performance of a particular foreign market, we will find it more interesting to compare the relative performance of a representative foreign *portfolio* with that of the U.S. equity market. To do this, we select a broadly diversified portfolio of the six stock markets in our study and compute the portfolio's return and risk during 1975 through mid-1988.

Held unhedged, this portfolio's mean return over the 13 ½-year period was 17.7 percent; the annualized standard deviation of its monthly return was 17.0 percent. Thus, the foreign portfolio proved to be slightly more volatile than the U.S. stock market but recorded a notably higher return. The foreign portfolio's return/risk ratio, 1.04, represents a significant performance improvement over the U.S. equity market.

When we substitute currency-hedged stocks in our foreign portfolio, we get substantially lower risk, with only a marginal give-back of return. The foreign portfolio's standard deviation fell by some 3.2 percentage points (19 percent), down to 13.8 percent from 17.0 percent. The mean return on our hedged portfolio dropped 0.4 percentage points to 17.3 percent.[8] Thus, the portfolio of currency-hedged foreign equities registered a return/risk ratio of 1.26, a 40 percent improvement over that recorded by U.S. equities and 21 percent better than that of a portfolio of unhedged foreign stocks.

Notice that, just as it did with the individual foreign stock markets, currency hedging significantly reduced the volatility of our foreign portfolio. In fact, the 19 percent drop in the portfolio's volatility of return attributable to the elimination of currency risk is about the same as the reduction in the average foreign market's volatility. This indicates that holding a well-diversified portfolio of foreign stocks was no substitute for currency hedging. In the language of modern portfolio theory, exchange rate risks were not diversifiable.[9] Why not? Dollar exchange rates are highly correlated. In other words, because the dollar generally rises or falls against all foreign currencies in concert, holding positions in many different foreign equity markets does not eliminate your currency risks, as it would if dollar exchange rates were substantially uncorrelated.

Systematic versus Unsystematic Risk

When we compare the investment results of the hedged foreign portfolio with those of the unhedged portfolio, one feature stands out: their rates of return are

[8] The hedged portfolio's return fell by 40 bp per year, even though the average foreign market's return fell by only 10 bp., because Japan represents the largest share (30%) of our sample portfolio. Japan was the only stock market to perform substantially better unhedged rather than hedged. We consider the hedged and unhedged portfolios' performances to be essentially the same (see Figure 1). Put simply, foreign investors who bought U.S. stocks enjoyed about the same rate of return whether they currency hedged or not.

[9] If it were possible to diversify away exchange-rate risks, the reduction in the portfolio's volatility accomplished by currency hedging would have been much less than the reduction in the average single foreign market's volatility.

essentially the same (17.7 percent hedged, compared with 17.3 percent unhedged), but the hedged portfolio's returns were considerably less volatile. This *seems* inconsistent with the basic idea that lower risk can ordinarily be secured only by accepting lower return.

The key here is that you should expect to earn a lower return only if you reduce the *systematic* risk of your portfolio. Systematic risks are those that must be borne by some investor. They cannot be *eliminated* by reshuffling assets among portfolios, only *reassigned.* Exchange-rate risks are a different matter. A U.S. investor holding Japanese securities bears exchange rate risk. A Japanese investor holding U.S. securities also bears exchange rate risk. But if they forward contract with each other—that is, the U.S. investor agrees to sell yen and buy dollars, while the Japanese investor agrees to do the opposite—*both* can eliminate their exchange rate risk. Thus, the exchange rate risk is not systematic, and there is no reason to expect bearing exchange rate risks to command a risk premium.[10] Consequently, there is no reason to believe that a hedged portfolio will underperform (or outperform) an unhedged portfolio.

Of course, this does not mean that in particular subperiods you will not do better or worse—sometimes much better or much worse—by choosing to hedge. In fact the returns to hedged versus unhedged portfolios have often diverged substantially. To illustrate, Figure 1 shows the differences in cumulative average returns between our hedged portfolio and the otherwise identical but unhedged portfolio of foreign stocks. As you can see, at times the unhedged portfolio has been far in the lean—almost 7 percent per year, in fact, by the end of 1978. At other times the hedged portfolio has been the better performer. For example, by early 1985 the hedged portfolio had outdistanced the unhedged portfolio by about 5 percent per year.[11] But taking our period as a whole—1975 through mid-1988—the return difference between the two portfolios is negligible. The evidence in Figure 1 is consistent with our view that hedged and unhedged portfolios enjoy about the same return *over the long run.* Over any given *short run,* however, their performance can deviate substantially in either direction.

[10] This is a gross simplication of a complex topic. The conditions under which currency positions carry a risk premium are discussed in detail in Lee R. Thomas III, *International Differences in Asset Demands: Theory and Implications for the Risk Premium,* Financial Strategies Group Discussion Paper Series, No. 2, Goldman, Sachs & Co., July 1988. For author perspective, read Fischer Black, *Universal Hedging, International Equity Strategies,* Goldman Sachs & Co., May 1989.

[11] Note that these are annualized *cumulative* return differences. So Figure 1 shows that in early 1985 the hedged portfolio had outperformed its unhedged counterpart by an average of 5 percent *per year* in every year since the beginning of our sample period. Similarly, from 1975 to 1978 the hedged portfolio had earned about 7 percent *per year* less than the unhedged portfolio.

Figure 1
Cumulative Performance Difference

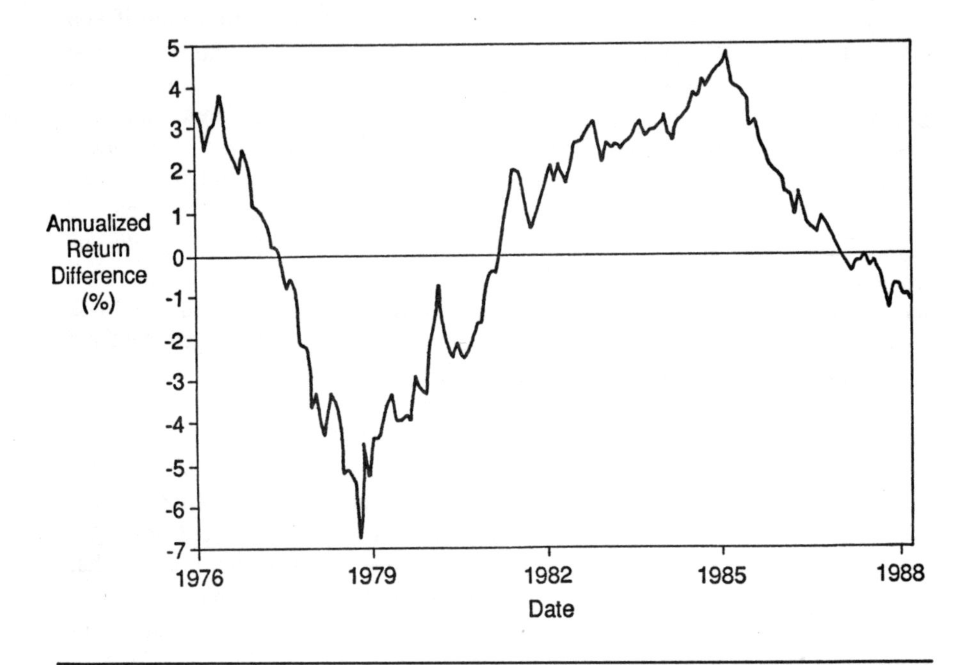

Foreign Equities as Diversification Vehicles

To gain further insights into the importance of international diversification, with and without currency hedging, we now look at the effects of diversifying into unhedged foreign stocks and compare our results with the effects of diversifying into hedged foreign stocks. Again, in both cases, our perspective is that of a U.S.-dollar-based investor.

Table 2 and Figure 2 illustrate the performance of portfolios containing *unhedged* foreign stocks. They show the average return and the volatility of a basket of stocks that combines the foreign equity portfolio described above with U.S. equities. The share allocated to foreign stocks varies from 100 percent (a portfolio of all foreign stocks) to 0 percent (a portfolio of all U.S. stocks) in 10 percent increments.

The results are broadly consistent with previous studies of international equity diversification. Adding foreign stocks to a U.S. portfolio almost invari-

Table 2 The Performance of Diversified Portfolios of U.S. and Unhedged Foreign Stocks (January 1975–June 1988)

Portfolio Shares		Unhedged		
% U.S.	% Foreign	Mean Return (%)	Std. Dev. (%)	Return/Risk Ratio
100	0	14.6	16.8	0.90
90	10	14.9	15.7	0.95
80	20	15.2	15.2	1.00
70	30	15.5	14.9	1.04
60	40	15.9	14.7	1.08
50	50	16.2	14.7	1.10
40	60	16.5	14.9	1.11
30	70	16.8	15.2	1.10
20	80	17.1	15.7	1.09
10	90	17.4	16.3	1.07
0	100	17.7	17.0	1.04

Figure 2 Unhedged Portfolio Performance

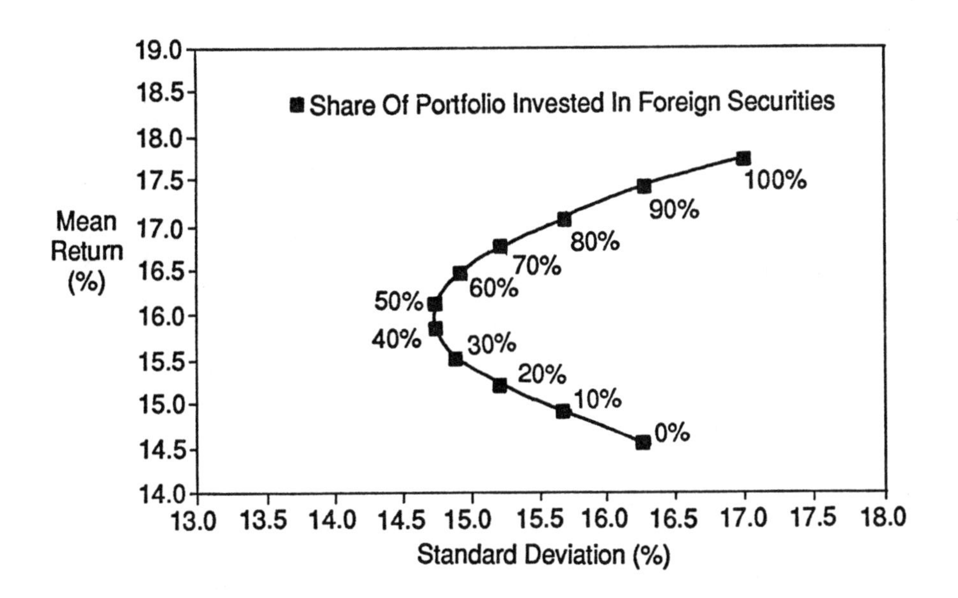

ably lowered its risk relative to a portfolio of U.S.-only stocks. In our study, the least risky portfolio contained about equal amounts of dollar and nondollar stocks. The riskiness of this portfolio, measured by its standard deviation, was 14.7 percent, compared with 16.3 percent for a portfolio consisting only of U.S. equities or 17.0 percent for the foreign-only portfolio. So, international diversification using unhedged foreign stocks offered a U.S. equity investor the opportunity to dampen the volatility of portfolio return by 1.6 percentage points, or about 10 percent.

With Table 3 and Figure 3, we see what happens when we diversify using hedged foreign equities. As before, the share of foreign assets rises from 0 percent to 100 percent in 10 percent increments.

With currency hedging, the minimum-risk portfolio contains substantially more foreign equities than the minimum-risk portfolio in the unhedged case--about 80 percent foreign and 20 percent U.S. stocks, compared with equal proportions of foreign and domestic stocks without hedging. The standard deviation of our minimum risk portfolio falls to 13.5 percent from 14.7 percent—a drop in volatility of 1.2 percentage points or roughly 8 percent. Thus, it appears that the risk reduction secured by hedging has been almost as great as that achieved by diversifying in the first place. And by combining diversification and currency hedging, we find that we can lower the standard deviation of our least risky portfolio to 13.5 percent—a 17 percent reduction in volatility compared with our original portfolio containing only U.S. stocks.

Improving the Hedge

We used an extremely simple hedge design for our investment simulations in the previous section: each month we sold forward the initial value of our foreign stock holdings for U.S. dollars. There are two potential problems with this hedging strategy that we may wish to address. First, because we hedged only the initial value of our foreign stock holdings, our monthly gains or losses (plus dividends) were currency exposed. Second, even though we based the hedge amount on the initial value of our foreign equity holdings, it is not clear that our hedge design—selling foreign currency forward pound-for-pound, yen-for-yen, or mark-for-mark to match our exposures—was best. In this section, we will examine the practical importance of these objections.

Hedging Gains or Losses

International portfolio managers often complain that currency hedging is difficult, since a perfect hedge requires selling forward the future foreign currency value of their foreign equity positions. This value, of course, cannot be known in advance. The hedge design we simulated in the previous section is realistic in

Table 3 The Performance of Diversified Portfolios of U.S. and Hedged Foreign Stocks (January 1975–June 1988)

Portfolio Shares		Unhedged		
% U.S.	% Foreign	Mean Return (%)	Std. Dev. (%)	Return/Risk Ratio
100	0	14.6	16.8	0.90
90	10	14.9	15.6	0.95
80	20	15.2	15.1	1.01
70	30	15.4	14.6	1.06
60	40	15.7	14.1	1.11
50	50	16.0	13.6	1.15
40	60	16.2	13.6	1.19
30	70	16.5	13.5	1.22
20	80	16.8	13.5	1.24
10	90	17.0	13.6	1.26
0	100	17.3	13.8	1.26

Figure 3 Hedged Portfolio Performance

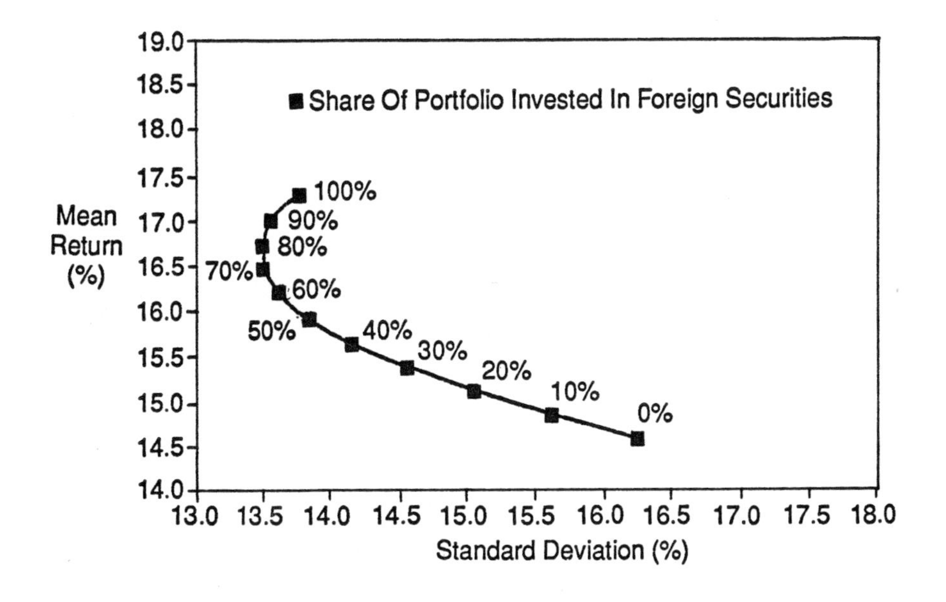

that it does not assume that you are prescient. It does, however, leave you currency exposed each month on your position's capital gains or losses and dividends. With this simple hedge, the unpredictable component of your return equals the product of the monthly exchange rate multiplied by the change in the value of the underlying stock position.

Let's take a look at what the impact on our portfolio would have been if we had *perfectly* hedged our currency risks, including our foreign capital gains or losses, over the entire 1975-mid-1988 period.

Table 4 compares stock market volatilities in four ways:

- measured in local currency terms,
- measured unhedged from a dollar investor's perspective,
- measured with our simple currency hedge in place, and
- measured using a hypothetical *ideal hedge.*

A bit of explanation is in order here as to what we mean by an "ideal hedge." This measure assumes that you can correctly forecast the foreign stock market's gain or loss each month, enabling you to hedge exactly the currency exposure embedded in the end-of-month value of your holdings. As you can see, the differences in the return volatilities recorded by the simple hedge design and our ideal hedge were small. On average, the annualized return volatilities of the two hedging strategies differed by only 0.22 percentage points. And in every market, both hedge strategies recorded volatility that was quite close to the volatility measured in local currency terms and much lower than the unhedged return volatility. That is, in practice, our simple, easy-to-implement hedging strategy

Table 4 Hedging Strategies Compared* (percent)

Market	Volatility in Local Currency Terms	Volatility Held Unhedged	Volatility Using a Simple Hedge	Volatility Using an Ideal Hedge
U.S.	16.3	—	—	—
Japan	14.7	20.4	14.8	14.6
Germany	17.2	21.0	17.4	17.2
U.K.	25.6	27.2	23.8	23.4
France	22.3	26.3	22.2	22.1
Switzerland	16.5	20.2	16.8	16.5
Canada	19.4	21.9	19.4	19.3

* Volatility is measured by the annualized standard deviation of monthly returns.

worked almost as well as an ideal hedge. In all probability, few international investors would notice the difference.

Picking a Hedge Ratio

Thus far, in our simulated trading, we have hedged an amount equal to the face value of the underlying foreign equity holdings, as we might if we were hedging a foreign bond. Actually, however, the exchange rate risks embedded in stocks are more subtle than those embedded in bonds. Let's now look at what these differences imply for designing a currency hedge.

Bonds are nominal assets; that is, bonds represent claims to cash. Foreign bonds are claims to foreign cash, and so they bear direct translation exchange rate risks. That is, bonds promise to pay you fixed amounts of foreign cash, which you must translate into your local currency at an uncertain exchange rate. With bonds, you can apply a simple currency-hedging strategy: you sell forward foreign currency with a present value equal to the present value of the bond's cashflows.[12] That is, after adjusting for the time value of money, you sell forward your bonds' future cash flows one-for-one: mark-for-mark, yen-for-yen, or pound-for-pound.[13]

The exchange rate risks embedded in equities are fundamentally different and much more complex than bonds' currency risks. Equity shares represent claims to real assets, not claims to fixed amounts of cash. As such, they do not bear simple translation risks.[14] This does not mean that equity prices are independent of changes in exchange rates, however. Equity shares are claims to real assets, but these are real assets specialized in the production of particular goods in particular countries. The *value* of assets specialized in the production of particular goods obviously depends on relative prices. Relative prices, in turn, de-

[12] Even hedging foreign bonds can be more complicated. If, our example, interest rates and exchange rates are correlated, the optimal hedge will involve selling more or less currency forward than the amount given by the bond. Fortunately, in practice, one-for-one foreign bond hedges appear to be adequate. (See Michael Adler and David Simon, "Exchange Risk Surprises in International Portfolios," *Journal of Portfolio Management*, Winter 1986.)

[13] This does not mean that you want to sell each of the foreign bonds' cash flows forward for its value on the date it is to be received. Such a hedge design would eliminate not only the bond's foreign exchange-rate risk, but also its foreign interest-rate risk. Instead you can use a rolling forward contract hedge design, which eliminates exchange-rate risk but leaves the bond's interest-rate risk in its original currency. (See Lee R. Thomas III, *Managing Currency Risks in International Bond Portfolios*, Sachs & Co., December 1987.)

[14] You *could* argue that the equity investor does face the problem of translation risk, because he needs to convert dividend payments and the future value of the foreign stock (upon its sale) from foreign currency into domestic currency. This reasoning is somewhat spurious, however, since the currency in which the company pays dividends—or the exchange where it is traded—is insignificant in terms of economics. That is, a Japanese company may agree to pay its dividends to U.S. investors in dollars instead of yen, evaluating the dollar value of the dividend at the spot exchange rate when the dividend is paid. But the exchange-rate risk of the company's stock will not be affected. Furthermore, you cannot avoid a stock's exchange-rate risks just by buying or selling shares in U.S. dollars. ADRs carry the same exchange-rate risks as their underlying foreign stocks.

pend on—or determine, depending on your view of the world—exchange rates. So, generally, the value of the assets that equity shares represent will be correlated with the level of exchange rates, particularly the level of real (inflation-adjusted) exchange rates.

Further, stock prices and exchange rates share common economic determinants. For example, both may change in response to monetary shocks. Once again, this means that changes in exchange rates and share prices will be correlated, although the magnitudes and even the signs of the correlations are difficult to predict before the fact on theoretical grounds alone.

Exchange-rate risks in foreign equities can also result from balance sheet or economic exposures specific to particular companies. A holder of a U.S. company's common stock, for example, may bear exposure to changes in the yen/dollar exchange rate if the company's primary competitor manufactures goods in Japan and imports them into the United States.

Formulating the Correct Hedge

In designing a hedge you should, in principle, estimate the sensitivity of your foreign equity holdings to exchange rate changes and use the resulting estimate to choose the correct hedge ratio. Before the fact, there is no reason to believe that the best hedge ratio will be 1.0—that is, that the ideal hedge for an equity holding generally will be mark-for-mark, yen-for-yen, or pound-for-pound.[15] So instead, let's specify that our hedging strategy will be to sell h units of foreign currency forward for each unit value of foreign equity held, where h is a hedge ratio estimated using the appropriate regression equation.

Before considering the effects of using an optimal hedge ratio in place of our simple hedge design, you should be aware that the simulation results that follow are bound to overstate the gain in effectiveness that you can expect to enjoy in practice from estimating a hedge ratio statistically. In practice, you won't know what the best hedge ratio is until after the fact. Your before-the-fact estimate will invariably contain some error. Our results implicitly assume that you could exactly deduce the correct hedge ratio to use in advance.

Table 5 shows the results of our attempts to estimate ideal hedge ratios. Four of the six estimated best hedge ratios are greater than 1.0, while two are less than 1.0. But as the second column of the table shows, only one of those greater than 1.0—Canada—is significantly different from 1.0 in a statistical sense. Neither of those less than 1.0 is significantly different from 1.0.

More important than the *statistical* significance of the estimated hedge ratios is the question of whether they produce significantly better hedging results in *economic* terms. The last column of Table 5 shows how much we were able to lower the volatility of return by using the statistical hedge ratios in place of

[15] The ideal hedge ratio is 1.0 only when exchange-rate changes and equity returns are uncorrelated.

Table 5 Statistical Hedge Ratios (January 1975–June 1988)

	Risk-Minimizing Hedge Ratio "h"	t-Value*	Reduction in Return Volatility Compared with a Simple One-to-One Hedge
Japan	1.16	1.71	0.1%
Germany	0.95	–0.72	0.0%
U.K.	1.12	–0.40	0.1%
France	1.21	1.42	0.1%
Switzerland	0.82	–1.83	0.1%
Canada	2.33	4.52	1.2%

* For a test of the hypothesis that "h" equals 1.0. Note that we can reject that hypothesis at the .05 significance level only in the case of Canada.

our simple hedge ratio—1.0. As you can see, in every case except Canada, the reduction in volatility is essentially insignificant. In the case of Canada the best hedge ratio, 2.33, is substantially—indeed, implausibly—larger than 1.0. Using this ratio permitted an investor to reduce the volatility of his stock holdings well below the local currency volatility; so even a Canadian domestic equity investor would have wanted to use a currency hedge. We view the result for Canada as somewhat suspect and would not predict that this hedge ratio will prove stable in future years.

Implications

Because different dollar foreign exchange rate changes are highly correlated, exchange rate risks in an internationally diversified portfolio are not self-hedging. Exchange-rate risks tend to accumulate rather than to diversify away, even if you hold stocks from many foreign markets. Thus, international diversification is *not* a substitute for currency hedging. Rather, currency hedging complements international equity portfolio diversification.

Based on data from 1975 through mid-1988, we estimated that the volatility of a U.S.-only equity portfolio was 16.3 percent. International diversification alone enabled us to reduce this by about 10 percent.[16] Currency hedging permitted a further 8 percent reduction, to 13.5 percent. Thus, currency hedging offered an equity portfolio manager almost the same amount of risk reduction as

[16] These results refer to risk-minimizing portfolios. Of course, before the fact portfolio managers could not have known the portfolio weights necessary to minimize risk. So this overstates the potential for reducing portfolio volatility by diversifying internationally. But since we deliberately chose a simple hedge design, the results do not necessarily overstate the potential for reducing volatility by hedging.

did diversifying abroad in the first place. Furthermore, this reduction in volatility did not require using sophisticated hedging instruments or strategies—a simple hedge design worked almost as well as two other, more complicated approaches we examined.

Our results do not imply that portfolio managers should always hedge. If you have strong views that particular foreign currencies will outperform their forward exchange rates, you may wish to express your beliefs by remaining unhedged or by partially hedging.[17] But if you have no strong exchange rate views—i.e., if you believe that forward rates are approximately unbiased predictors of future spot rates—you can reduce your portfolio's volatility by hedging, without reducing its expected return. In this case, if volatility is important to you, you should hedge.

There are many misapprehensions about managing currency risks in international portfolios. Some practitioners object to the very idea of currency hedging, because they believe that the objective of diversifying into foreign equities is to secure currency exposures. This is unreasonable. If you want currency exposures in your portfolio, the best way to get them is by adding pure currency investments, such as foreign currency deposits, not foreign stocks. Stocks bear equity risks that are extraneous if all you want is currency exposure. Tactically you should add foreign stocks to your portfolio when you expect foreign equity markets to outperform U.S. markets when you compare them in a currency-neutral way.

Strategically, you should add foreign stocks to diversify your domestic equity risks. If you also happen to want complementary currency exposures—and the amount of currency exposure you want to bear happens to conform, coincidentally, to the amount of foreign stocks you want to hold—then a policy of neglecting the currency composition of your portfolio makes sense. Otherwise the best strategy in any period will involve some action: to double up your currency exposures, to partially hedge them, or to overhedge—but never to ignore them.

The point is, in the end whether you hedge or not, as an active manager you should always evaluate the currency composition of your portfolio and consider hedging. Not hedging is an active decision. Whichever path you choose, you should subsequently assess the investment results of your net currency exposures independently of the market or individual asset choices you made. And in either case, you should always consider the effects of your currency decisions on both return and risk.

[17] Completely passive investors may also choose not to hedge, reasoning as follows: if foreign currency exposures carry a risk premium, I am exactly adequately compensated for bearing the risk; if not, it is because the risk is not systematic and can therefore be safely ignored.

CHAPTER 8

Introduction to International Bonds

Robert A. Brown
Ibbotson Associates, Inc.

Laurence B. Siegel
Ibbotson Associates, Inc.

A Whirlwind of Change

Bond markets around the world have been experiencing rapid change. This change has occurred in many dimensions: market capitalization, diversity and complexity of instruments, trading volume, the number of countries participating, and the increasingly global character of transactions and holdings.

In total, the world bond market has grown from $700 billion in 1966 to $11 trillion today, a growth rate of 13 percent per year. The U.S. bond market has always composed a large share of world bond market capitalization. From approximately $300 billion in 1966, it has grown to $4.9 trillion today, or about 45 percent of the world bond market.

Figure 1 shows the composition of the world bond market by type of bond. Government bonds are the largest component. Corporate bonds make up a lesser share, and cross-border bonds are the smallest but fastest growing. We now address each of these categories.

Government Bonds

The two primary segments of the global bond market are governments and corporates. Government bonds have been the faster growing segment. In 1967 their

Figure 1 Global Distribution of Government, Corporate, and Cross-Border Bonds

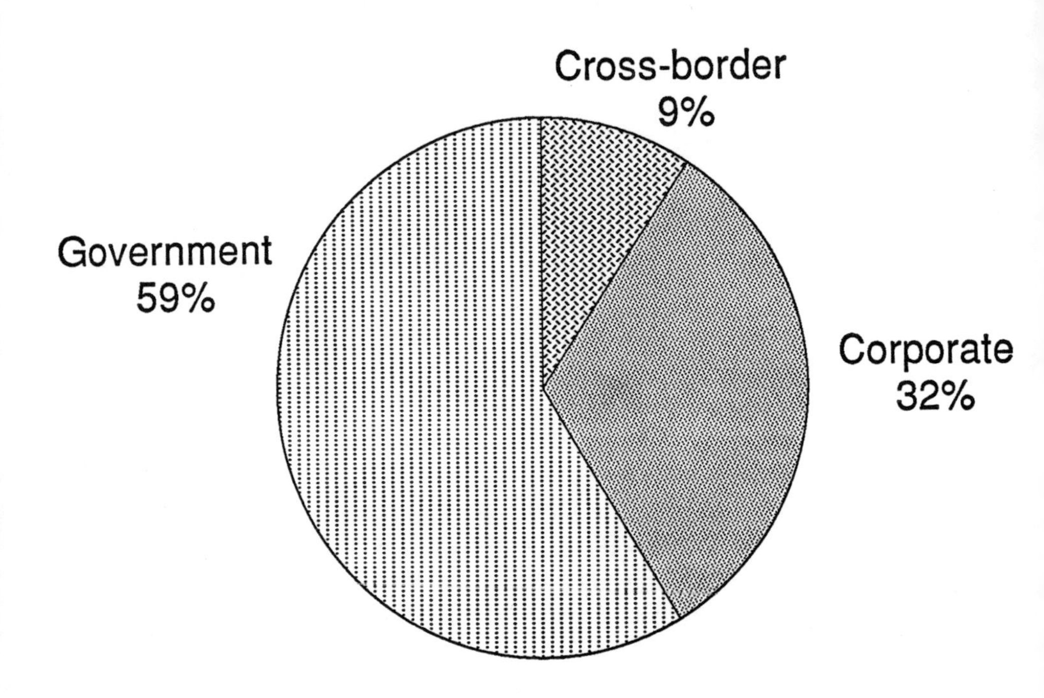

worldwide market value from around the world summed to $400 billion. Today this value is $6.5 trillion, representing a 14 percent annual growth rate. Figure 2 presents the current distribution of the global market for government bonds. This segment is currently dominated by the United States, with significant participation by Japan, Germany, France, and the United Kingdom. Interestingly, Japan was not a leading government bond issuer until relatively recently; Japan was constitutionally forbidden to issue debt until 1966.

Figure 2 Global Distribution of Government Bonds

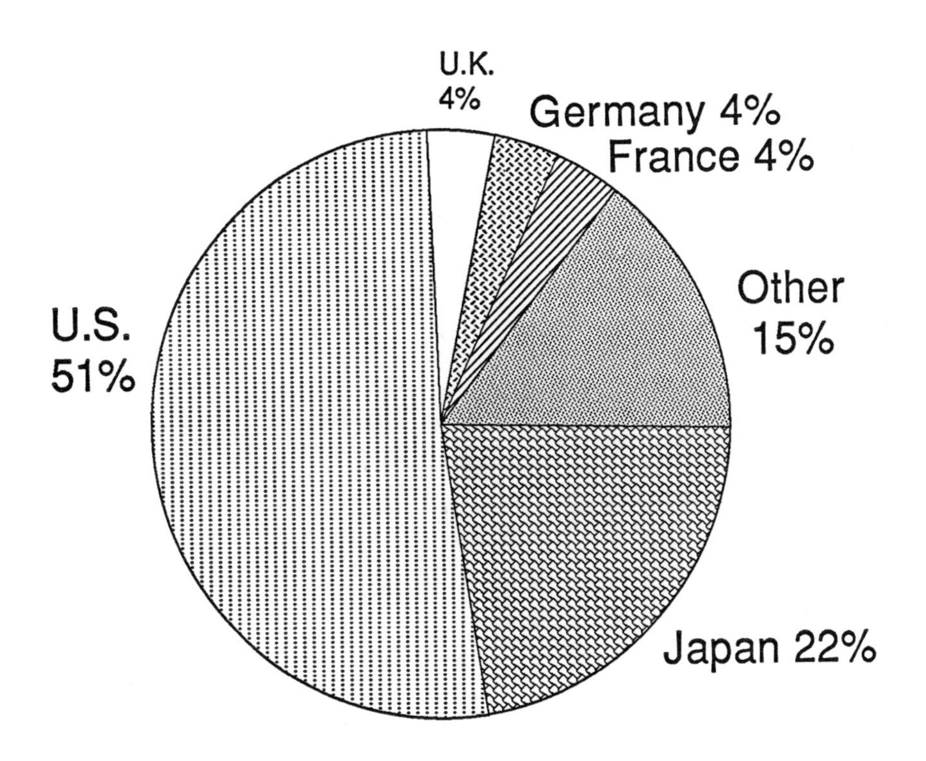

Corporate Bonds

The market for corporate bonds, although not as large as that for governments, is still rapidly growing and of great importance. In 1966 corporate bonds composed 37 percent of the world bond market. Since then this market has grown at a 10 percent annual rate, slower than governments, so that today corporate bonds occupy only 32 percent of the global market. Figure 3 shows the distribution by country for the major participants in this market. Today, this market is dominated by Japan, the United States, and Germany.

Figure 3 Global Distribution of Corporate Bonds

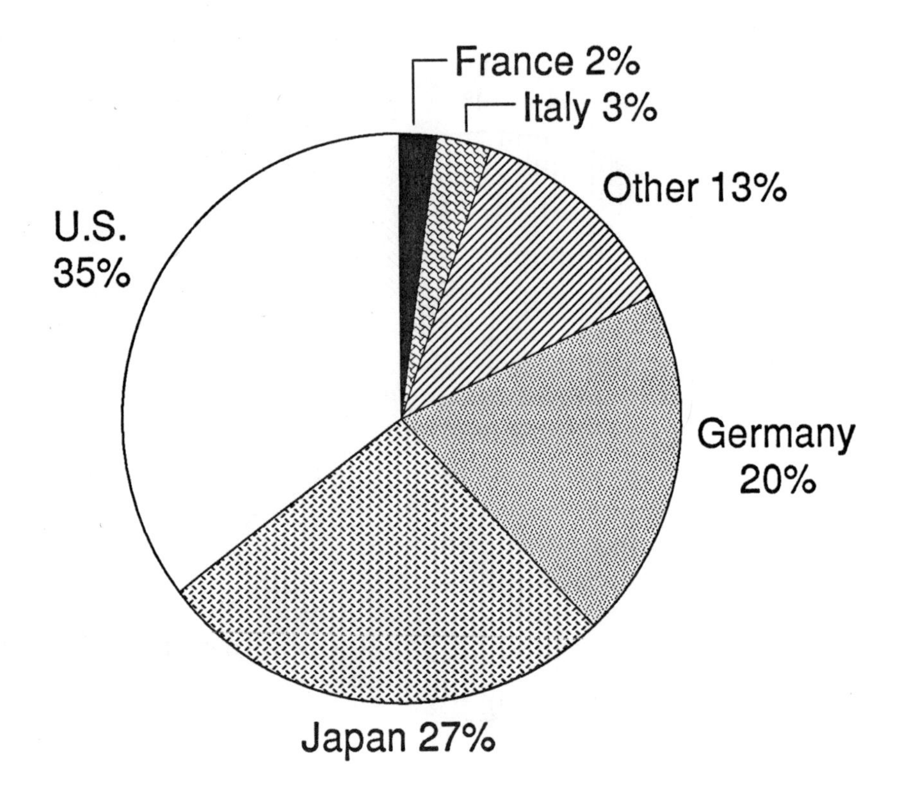

The well-publicized mountain of corporate debt, then, is overshadowed by government debt. Debt has always been considered a necessary and prudent part of the capital structure of corporations. Government indebtedness, however, has historically been motivated by wartime needs. The current high level of worldwide government debt in peacetime reflects public tolerance of such indebtedness and demand for services provided by governments.

Cross-Border Bonds

A rapidly developing subset of the government and corporate sectors is the market for *cross-border* bonds. Cross-border bonds are bonds issued by a country or one of its corporations, in the currency of another country and/or primarily held by residents of another country. Cross-border bonds include traditional foreign bonds, and various categories of Eurobonds. (There is nothing uniquely European about Eurobonds except their place of origin nearly three decades ago; Eurobonds are traded worldwide.) In 1967 the cross-border bond market accounted for only $27 billion. Today it has reached $945 billion, having grown at an 18 percent annual rate.

The Eurobond market arose in the early 1960s in response to the imposition of the Interest Equalization Tax by the United States. Interestingly, with the repeal of the tax a decade later, this market did not die out but became even more prominent. The development of the Eurobond market, which is relatively free of government controls, has aided the free flow of capital among nations.

Opportunities in International Bond Management

The worldwide growth in corporate, government, and cross-border bonds has significantly expanded the opportunities available to U.S. fixed-income managers. This is true not just of the giant European and Japanese markets, but of vibrant smaller markets such as Australia and New Zealand.

Moreover, the need for an international perspective has been heightened by the shrinking portion of the world bond market occupied by U.S. issues. It is reasonable to believe that bond market growth outside the United States will continue to outstrip domestic bond growth to such a degree that internationally diversified bond portfolios will become the norm in the future.

Historical Returns on Government Bonds

Historically, local-currency returns of the various bond markets around the globe have been quite uniform, with a few exceptions, in terms of their summary statistics. Table 1 shows summary statistics of the local-currency and U.S. dollar returns on major long-term government bond markets over roughly the last three decades. (Long-term bonds are generally those with 10 years and longer to maturity.) In local currency, only Switzerland has a compound annual return outside the range of 6 to 9 percent.

Table 1 Summary Statistics of Annual Returns on Long-Term Government Bonds

		Total Returns in Local Currency			Total Returns in U.S. Dollars		
Country	Period	Compound Annual Return (%)	Arithmetic Mean Return (%)	Standard Deviation of Return (%)	Compound Annual Return (%)	Arithmetic Mean Return (%)	Standard Deviation of Return (%)
Australia	1961–89	7.12	7.69	11.19	5.84	6.56	12.72
Canada	1961–89	7.64	7.94	8.58	7.08	7.42	8.99
France	1961–89	8.00	8.39	9.33	7.40	8.41	15.51
Germany	1961–89	7.09	7.31	7.01	10.48	11.34	14.25
Italy	1961–89	8.80	9.88	15.95	6.16	8.04	21.59
Japan	1967–89	7.58	7.76	6.37	11.99	13.33	17.80
Netherlands	1965–89	7.24	7.46	7.21	9.99	10.92	14.97
Switzerland	1965–89	4.33	4.51	6.28	8.72	9.93	16.85
United Kingdom	1961–89	8.80	9.83	15.97	6.75	8.51	21.07
United States	1961–89	6.18	6.70	11.17	6.18	6.70	11.17

Translated to U.S. dollars, these returns are quite different across countries because of exchange rate fluctuations. Japanese bonds returned 12 percent per year in U.S. dollar terms, while Australian bonds returned less than 6 percent. The standard deviation (risk) of bonds measured in U.S. dollar terms was also higher than when measured in local currency. This observation highlights the importance of exchange risk in global bond management.

Returns in Local Currency

Focusing on the United States, government bonds had a geometric mean return of 6.2 percent, the second lowest local-currency return of the major markets. Moreover, the U.S. standard deviation was 11.2, higher than most other countries.

The local currency returns shown in Table 1 provide an approximate indication of what would have occurred for a U.S. investor with a currency-hedged position. A hypothetical perfect hedge would provide the local currency return; in practice, no hedge is perfect or costless, and the U.S. investor's results would differ from the local-currency return shown.

Returns in U.S. Dollars

The unhedged U.S. investor earned the U.S. dollar-translated returns shown in Table 1. Again, the U.S. market exhibits one of the lowest average returns. Now, however, U.S. volatility is lower than that of any other country, with the exception of Canada. Exchange rate fluctuation is responsible for this turnabout.

Not surprisingly, Japan and Germany had the highest returns. This was entirely the result of appreciation of the Japanese yen and German mark against the dollar during this time period, since these two countries had only average returns measured in local currency.

Gains from Diversification

One of the principal advantages of a global approach to bond management is the reduction of portfolio risk through diversification. Diversification benefits can be seen in the cross-correlation coefficients between countries. Tables 2 and 3 provide these correlations in local currency and U.S. dollars, respectively. The bottom row in each of these two tables shows the correlation between the U.S. long-term government bond market and each of the foreign markets. The correlation coefficients of the United States with the foreign markets, except Canada, are quite low. This indicates significant risk reduction potential through international diversification.

Another way of quantifying the benefits of globally diversifying a government bond portfolio is to compare returns on diversified and undiversified port-

Table 2 Correlation Coefficients of Long-Term Government Bonds in Local Currency, Based on Annual Data 1961–89

	1	2	3	4	5	6	7	8	9	10
1. Australia	1.00									
2. Canada	.50	1.00								
3. France	.58	.63	1.00							
4. Germany	.40	.49	.51	1.00						
5. Italy	.43	.41	.72	.32	1.00					
6. Japan	.34	.19	.45	.61	.29	1.00				
7. Netherlands	.53	.76	.60	.79	.37	.50	1.00			
8. Switzerland	.40	.44	.51	.73	.28	.55	.65	1.00		
9. United Kingdom	.51	.49	.44	.58	.34	.50	.60	.54	1.00	
10. United States	.52	.90	.63	.45	.40	.17	.70	.43	.38	1.00

Table 3 Correlation Coefficients of Long-Term Government in U.S. Dollars, Based on Annual Data 1961–89

	1	2	3	4	5	6	7	8	9	10
1. Australia	1.00									
2. Canada	.16	1.00								
3. France	.31	.21	1.00							
4. Germany	.12	.16	.68	1.00						
5. Italy	.39	.16	.72	.45	1.00					
6. Japan	.33	-.09	.56	.60	.48	1.00				
7. Netherlands	.14	.31	.77	.93	.50	.54	1.00			
8. Switzerland	.13	-.06	.67	.89	.36	.69	.86	1.00		
9. United Kingdom	.39	.12	.38	.41	.30	.62	.42	.40	1.00	
10. United States	.07	.81	.46	.36	.35	.05	.50	.15	.16	1.00

folios. The first section of Table 4 compares long-term U.S. government bonds with an equally weighted global portfolio of like maturity. The global portfolio has a higher compound annual return and lower standard deviation, both in local currency (roughly speaking, hedged) and U.S. dollar (unhedged) terms.

Table 4 Gains from International Diversification: U.S. versus Global Portfolios (All Returns in U.S. Dollars)

Portfolio	Period	Compound Annual Return (%)	Arithmetic Mean Return (%)	Standard Deviation of Return (%)
Government				
Long-Term U.S. Government Bonds	1960–89	6.42	6.94	11.06
World Government Bonds[1]	1960–89	8.23	8.50	7.86
Corporate				
U.S. Corporate Bonds	1960–89	6.88	7.41	11.29
Global Corporate[2]	1960–89	9.59	9.93	9.18
Cross-Border				
U.S. Corporate Bonds	1978–89	10.34	11.21	14.93
Cross-Border Bonds[3]	1978–89	10.43	11.16	13.56

1 Value-weighted aggregation of foreign and U.S. government bonds.
2 Value-weighted aggregation of foreign and U.S. corporate bonds.
3 Equally weighted aggregation of Eurodollar, Euro-Sterling, Euro-Mark, Euro-Yen, and Euro-Franc bonds.

Figure 4 Growth of $1 Invested at Year-End 1959 in Long-Term Government Bonds

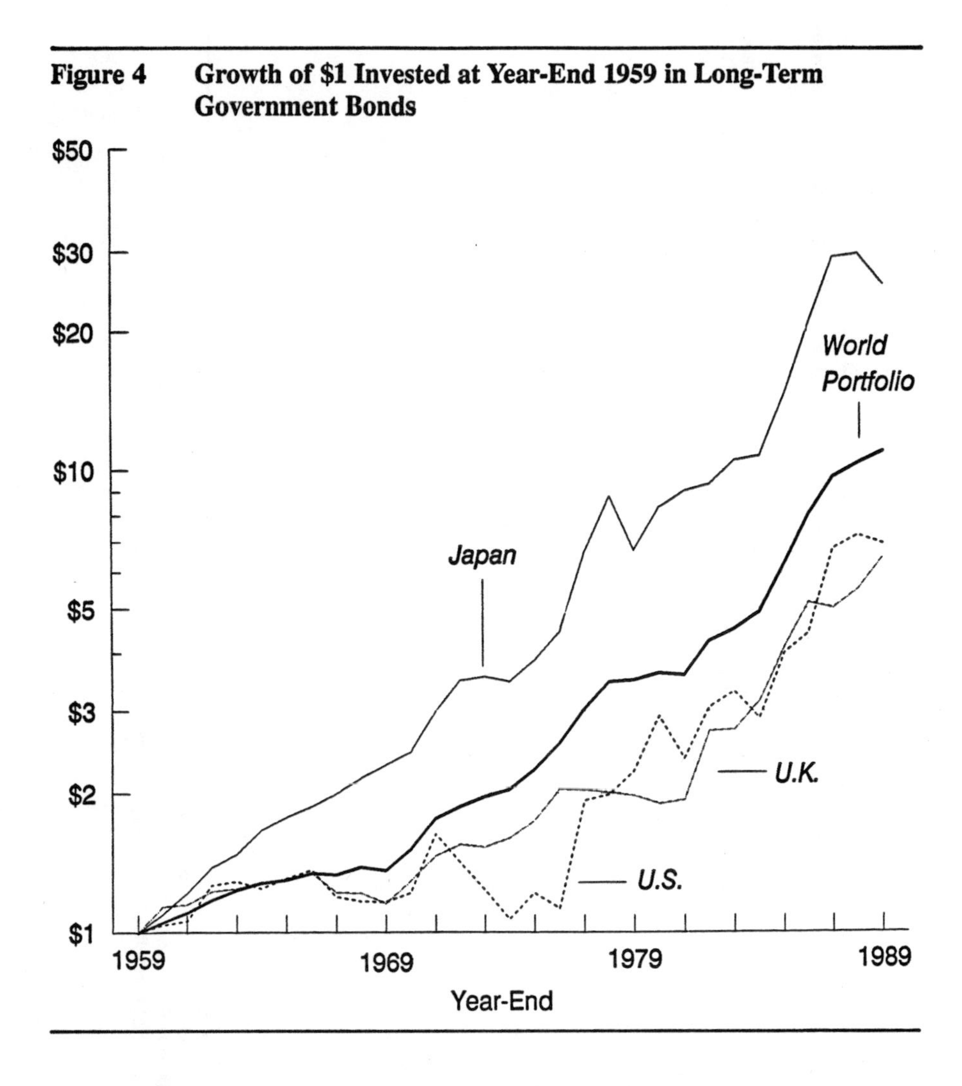

Figure 4 graphically presents these results. One U.S. dollar is invested at the end of 1959 in U.S. and global bond portfolios. The world portfolio beat the U.S. portfolio substantially, with a dollar growing to $10.95 over the 30-year period, compared to $6.48 for U.S. bonds alone.

Historical Returns on Corporate Bonds

Corporate bond returns are closely related to government bond returns because they are both driven by the general level of interest rates in a given country.

Corporates differ from governments in that they have higher yields. Part of this yield spread is expected to be consumed by defaults, and the remainder is a true risk premium captured by the investor as compensation for exposure to default risk.

Table 5 shows summary statistics for corporate bonds in the primary markets. The figures on the left-hand side are presented in local currency terms and are, therefore, comparable to a hedged position, ignoring hedging costs and imperfections.

As with government bonds, U.S. corporates provided one of the lowest rates of return while at the same time exhibiting a middle level of volatility. The right-hand side of Table 5 shows returns on the same bonds expressed in U.S. dollars. With the influence of fluctuating exchange rates, volatility of the U.S. market improves to best (lowest) place. However, the average return remains one of the lowest.

The diversification benefits of international corporate bonds are seen in Tables 6 and 7. These tables present the correlation coefficients for corporate bonds in local currency and U.S. dollars, respectively. The bottom row in each table shows correlations of the U.S. market with each foreign market; these generally low correlations suggest the potential for diversification benefits.

Referring back to Table 4, the second section of the table quantifies these benefits. The "global corporate" portfolio is composed of equal portions of corporate bonds from the U.S., Canada, France, Germany, and Japan. This portfolio provided an average return 2.7 percent per year higher than U.S. corporate bonds alone. At the same time, the standard deviation of the global portfolio was 0.6 percent lower.

Historical Returns for Cross-Border Bonds

The fastest growing segment of the international bond market has been the market for cross-border bonds. This market consists largely of issues sold in one country and currency by a borrower of a different nationality; also, the country (primary venue of sale) and currency may not match. Categories of cross-border bonds include Eurodollar and other Eurocurrency (e.g., Euro-Mark and Euro-Yen) bonds; Yankee, Samurai, and Bulldog bonds; and traditional foreign bond markets.

Table 8 shows the historical returns of various cross-border categories measured in local currencies and U.S. dollars. In local currency, bonds denominated in French francs had the highest return over the 1978–1989 period; dollar-denominated bonds were the most volatile. Translated to U.S. dollars, cross-border bond returns were very similar to each other across categories; bonds denominated in yen had the highest return due to the appreciation of the yen against the dollar over the period. As before, when returns are translated to U.S. dollars, the

Table 5 Summary Statistics of Annual Returns on Corporate Bonds

		Total Returns in Local Currency			Total Returns in U.S. Dollars		
Country	Period	Compound Annual Return (%)	Arithmetic Mean Return (%)	Standard Deviation of Return (%)	Compound Annual Return (%)	Arithmetic Mean Return (%)	Standard Deviation of Return (%)
Austria	1965–86	7.86	8.00	5.79	11.02	11.76	13.44
Canada	1961–86	7.92	8.24	8.70	6.57	6.87	8.28
France	1961–86	9.16	9.60	10.12	8.01	9.01	15.77
Germany	1961–86	8.72	9.03	8.97	11.97	12.55	11.80
Italy	1961–86	10.35	11.46	16.16	7.08	8.96	21.44
Japan	1961–86	8.49	8.65	6.08	11.93	12.68	13.52
Netherlands	1961–86	6.71	6.98	7.89	8.96	9.67	12.96
United Kingdom	1961–86	9.59	10.53	15.08	6.92	8.21	17.50
United States	1961–86	6.65	7.04	9.48	6.65	7.05	9.48

Table 6 Correlation Coefficients of Corporate Bonds in Local Currency, Based on Annual Data 1961–86

	1	2	3	4	5	6	7	8	9
1. Australia	1.00								
2. Canada	.71	1.00							
3. France	.81	.74	1.00						
4. Germany	.28	.46	.50	1.00					
5. Italy	.55	.55	.77	.49	1.00				
6. Japan	.20	.15	.08	.16	.14	1.00			
7. Netherlands	.55	.74	.60	.58	.39	.24	1.00		
8. United Kingdom	.69	.66	.56	.36	.53	.40	.65	1.00	
9. United States	.70	.91	.72	.44	.52	.21	.72	.67	1.00

Table 7 Correlation Coefficients of Corporate Bonds in U.S. Dollars, Based on Annual Data 1961–86

	1	2	3	4	5	6	7	8	9
1. Australia	1.00								
2. Canada	.34	1.00							
3. France	.79	.31	1.00						
4. Germany	.79	.33	.67	1.00					
5. Italy	.51	.30	.76	.49	1.00				
6. Japan	.48	.20	.56	.40	.50	1.00			
7. Netherlands	.88	.48	.75	.81	.47	.45	1.00		
8. United Kingdom	.45	.29	.50	.27	.46	.54	.42	1.00	
9. United States	.39	.88	.47	.39	.38	.26	.53	.41	1.00

currency effect boosts the volatility of foreign-currency issues beyond that of U.S. dollar-denominated (Eurodollar and Yankee) issues.

The correlations of issues in different currencies are shown in Table 9 and indicate gains from diversification. The third section of Table 4 shows results for a hypothetical equally weighted portfolio of cross-border bonds. The "cross-bor-

Table 8 Summary Statistics of Annual Returns on Cross-Border Bonds

		Total Returns in Local Currency			Total Returns in U.S. Dollars		
Currency	Period	Compound Annual Return (%)	Arithmetic Mean Return (%)	Standard Deviation of Return (%)	Compound Annual Return (%)	Arithmetic Mean Return (%)	Standard Deviation of Return (%)
Euro C$	78–89	10.65	10.80	5.98	10.13	10.34	7.00
Euro-Franc	78–89	11.62	11.88	8.00	9.73	11.08	18.31
Euro-Mark	78–89	6.72	6.86	5.69	8.69	10.10	18.77
Samurai	78–89	6.92	7.02	4.89	11.58	13.41	20.82
Euro-Yen	78–89	6.77	6.90	5.36	11.43	13.27	20.87
Foreign Guilder	78–89	8.41	8.57	6.22	10.03	11.34	18.17
Euro-Guilder	78–89	8.21	8.33	5.40	9.83	11.10	17.88
Foreign SwFr	78–89	3.90	4.01	5.04	6.18	7.71	19.49
Euro-Sterling	78–89	10.96	11.23	8.17	9.44	10.68	17.71
Yankee	78–89	10.86	11.50	12.79	10.86	11.50	12.79
Eurodollar	78–89	10.37	10.68	8.89	10.37	10.68	8.89
U.S. Gov't	78–89	10.15	10.74	12.22	10.15	10.74	12.22

C$ – Canadian dollar.
SwFr – Swiss franc.
U.S. Gov't – Domestic U.S. government bonds for comparison.

Table 9 Correlation Coefficients of Cross-Border Bonds in U.S. Dollars, Based on Annual Data 1961–86

	1	2	3	4	5	6	7	8	9	10	11
1. Euro-C$	1.00										
2. Euro-Franc	-.18	1.00									
3. Euro-Mark	-.10	.94	1.00								
4. Samurai	-.14	.67	.64	1.00							
5. Euro-Yen	-.12	.67	.67	.99	1.00						
6. Foreign Guilder	-.14	.93	.99	.63	.66	1.00					
7. Euro-Guilder	-.16	.93	.99	.65	.67	.99	1.00				
8. Foreign SwFr	-.25	.88	.94	.75	.78	.93	.94	1.00			
9. Euro-Sterling	-.08	.62	.69	.57	.56	.71	.73	.66	1.00		
10. Yankee	.47	.30	.33	.05	.06	.36	.32	.10	.06	1.00	
11. Eurodollar	.49	.18	.22	.01	.01	.25	.21	-.01	-.04	.99	1.00

C$ – Canadian dollar.
SwFr – Swiss franc.

der portfolio" is composed of equal portions of Eurodollar, Euro-Sterling, Euro-Mark, Euro-Yen, and Euro-Franc bonds. This portfolio outperformed a U.S. corporate bond portfolio of similar maturity and quality by 0.1 percent compounded over 1978–1989 with a 1.4 percent reduction in the standard deviation of annual returns.

Issues in Global Bond Management

Currency Hedging

Investing in bonds denominated in foreign currencies raises the question of hedging. The case for hedging is usually made as follows: currency exposure has risk but no expected return, so it should be avoided. The price of avoiding currency risk includes, but is not limited to, the direct cost of the hedge. Existing instruments do not allow perfect hedging; some currency risk will remain in a hedged portfolio. The size and nature of these hedging imperfections determines the desirability of hedging.

On the other hand, it is not necessarily true that currency exposure has no expected return. The expected return to currency exposure can be positive to one party and negative to the counterparty, because the two parties are not taking the same risk. The relevant risk is not the standard deviation of currency fluctuations (which is the same for both parties) but the effect of these fluctuations on each party's portfolio. This effect differs according to the correlation of the currency with the assets in the portfolio.

There are other choices besides fully hedged and unhedged positions. A portfolio may be partly hedged. Another choice is to use *currency insurance*, which works like portfolio insurance: the exposure to the risky asset (in this case, the foreign currency) is increased, the higher the current position above a prespecified floor. Like portfolio insurance, currency insurance has a cost associated with guaranteeing a minimum return on the currency.

The currency hedging problem, then, is determined by clientele effects.[1] An investor whose other assets have a lower correlation with a hedged global bond portfolio than with an unhedged global bond portfolio should hedge. (This assumes that the expected return does not change after accounting for the cost of the hedge.) An investor dedicating foreign bonds to pay foreign liabilities in the same currency should not hedge. An investor willing to pay additional costs in good times in order to avoid adverse currency movements should purchase currency insurance. In fact all of these positions are held. It should thus be pre-

[1] Empirical studies showing a benefit to hedging, or not hedging, should be regarded with caution. Since fixed exchange rates were not abolished until in the early 1970s, we believe that there are not enough years of data to test any general hypothesis about the desirability of currency hedging

sumed, in the absence of compelling evidence to the contrary, that all of these are fairly priced, i.e., attractively priced to one clientele or another.

Passive Global Management

The effectiveness of passive management of equity portfolios has sparked interest in a parallel approach to bonds. In equities, passive management refers largely to indexing—that is, construction of a portfolio with the explicit intent to match a particular index. Such equity indices are typically market-capitalization weighted and broadly diversified.

Indexing in the literal sense does not necessarily travel well to the bond market. Most investors have a (more or less diffuse) planning horizon, causing them to prefer long, intermediate, or short-term bonds. A portfolio that is diversified across maturities in proportion to the market capitalization of these maturities is likely to be suboptimal for many investors. This is true of global as well as domestic bond portfolios.

Market capitalization weights may also mislead the investor seeking to diversify a bond portfolio across countries or currencies. The capitalization of a country's bond market may largely be an artifact of the country's institutional peculiarities. An example is the extremely small size of the British publicly traded corporate bond market. It gives no indication of the great importance of British corporations in the world economy.

Passive bond management, then, has come to mean more than indexing. Country weights may be set according to criteria other than market capitalization: gross domestic product (GDP) weights, which capture the importance of a country's whole economy, are one choice. Equal weights and stock market capitalization weights are other possible choices. An interesting passive country-weighting system, based on liquidity, is used in the J. P. Morgan Global Government Bond Indices. Their "benchmark" portfolio is based on a very few extremely liquid (on-the-run) bonds; in Japan, for example, one issue is used. The "active" portfolio consists of the benchmark bonds plus other heavily traded bonds. The "traded" portfolio is composed of the active bonds plus other traded bonds. Finally, a comprehensive portfolio includes the traded bonds plus less liquid bonds. All of the J. P. Morgan portfolios are weighted according to the market capitalizations of the particular bond issues in the various indices.

Because different investors have different bond horizon preferences, sub-indices by maturity or duration may also be useful to passive global bond managers. Such portfolios are sometimes difficult to fill, because not all parts of the yield curve are adequately populated in all countries. Nevertheless, an internationally diversified portfolio of bonds targeted to a particular time horizon (maturity or duration) is likely to be desirable to many investors.

CHAPTER 9

A Practitioner's Perspective of Managing Futures Within an Australian Fixed-Interest Portfolio

Peter Vann*
Westpac Investment Management Pty. Ltd.

In this chapter we consider the role of futures contracts in adjusting the return of an Australian fixed-interest portfolio. The three measures of a portfolio's return sensitivity used are internal rate of return, modified duration, and convexity. The theoretical basis for adjusting these return measures is discussed and then numerous examples presented for a number of parallel and nonparallel yield curve changes.

The liquidity, transaction cost, and parcel size differences are presented and the affect of these on portfolio adjustments discussed. We conclude that futures as an adjunct to physicals provide a mechanism for fine-tuning portfolio adjustments. The examples presented are for an index-tracking management approach but can also be applied to more active strategies, particularly where performance is measured against a performance benchmark such as a bond index.

We also compare the return sensitivity measures for the pricing basket of bonds from which the futures closeout yield is determined. We note that the modified duration and convexity for the pricing baskets differ from those of the futures. The mechanics of a squeeze at expiration of the futures is discussed, and

* Dr. Vann was formerly at Midland Montagu Australia Ltd., where this chapter was written.

we conclude that the recent moves by the Sydney Futures Exchange to a pricing basket of five bonds and the introduction of a deliverable contract will help minimize the possibility of squeezes at expiry.

Introduction

In this chapter we focus our attention on the use of futures on fixed-interest securities to assist in the management of a portfolio of fixed-interest securities.

If the futures are trading at fair value they provide the following attractions:

- Good liquidity enabling quick exposure to the market enabling timely modification of a portfolio's structure.
- Lower transaction costs than the physicals, including bid/ask spreads.
- Ability to trade in smaller units of face value.

In addition, if mispricing of the futures contract occurs, we may be able to take advantage of this to add value to a portfolio.

The major disadvantage is that for the bond market we have only two contracts with different term to maturities trading on the Sydney Futures Exchange (SFE). Thus, if we need to modify our portfolio in, for example, the 5- to 8-year range by using the 3- and 10-year contracts, we would be exposed to the risk of nonuniform yield curve movements (e.g., humping).

Under a given set of guidelines for managing a fixed interest portfolio, there will be set particular performance objectives. To enable us to focus our attention on the techniques, we will consider one management strategy for the examples given, index tracking. When tracking an index, we know exactly the adjustments required to our portfolio characteristics to enable the portfolio to track the index.

The index-tracking example used is a *passive* management style. An active management style commonly used is the benchmark approach. The benchmark is usually a market index, and the objective is to outperform the index. The enhanced returns required to outperform the benchmark index are obtained by correctly forecasting market movements, then restructuring the portfolio to provide higher returns if our views are correct. If we are taking short-term views, futures enable us to quickly set and then reverse the exposure we require. Typical gains made are between 5 to 20 basis points. The lower transaction cost and greater liquidity of futures ensures that we minimize our expenses.

The techniques discussed in the index-tracking examples can also be used for other strategies, and we will provide discussion on the benchmark management approach through the chapter. When we tilt our portfolio to gain a different market exposure than the benchmark index, we must also be aware of the risk of underperforming the benchmark.[1]

In the following sections of this chapter we will:

- Outline the characteristics of the current SFE contracts.
- Provide examples in the use of futures or physicals in rebalancing a portfolio.
- Consider the related issues of trading the bonds in the futures pricing basket and the squeeze at expiry.

SFE Bond Futures Contracts

The SFE currently has two bond contracts, a 3- and a 10-year contract, each contract being for $100,000 face value. Both are cash settlement contracts at expiry based on the price of a hypothetical 12 percent coupon bond whose term to maturity is 3 or 10 years on expiry of the contract. The yield used for cash settlement at expiry is the average yield of a basket of bonds. The composition of the baskets for the December 1989 contracts are in Table 1. The closeout yields are obtained from a number of brokers, the lowest and highest two numbers are rejected and the rest averaged to determine the final yield. The contracts expires on the 15th of March, June, September, and December each year. The daily volume (in number of contracts per day) is shown in Figure 1. The bond futures provides the majority (51 percent excluding traded options) of the turnover at SFE.

There has been some concern with manipulation of the market near closeout of these contracts, and the SFE has announced its intention of introducing deliverable bond contracts in the near future. Holders of short futures positions close to expiry have three choices:

1. They can closeout before expiry by buying contracts.
2. Closeout their positions and simultaneously sell contracts for the next expiry, i.e., rollover.
3. Exercise their right to deliver an issue from the range of eligible deliverables. The short position holder then receives payment for the issue delivered.

[1] This will be discussed in a forthcoming article "Definition and Control of Risk Measures for Active Portfolio Management and the Role of Options," to be published in 1990, by Dow Jones.

Table 1 Pricing Stocks for December 1989 Contracts

	10-Year Bond Futures	
12.5%	March	1997
12.5%	September	1997
12.5%	January	1998
12.0%	July	1999
13.0%	July	2000
	3-Year Bond Futures	
13.0%	February	1992
12.0%	March	1992
12.5%	July	1992
13.0%	May	1993

Figure 1 Daily Turnover SFE Bond Futures

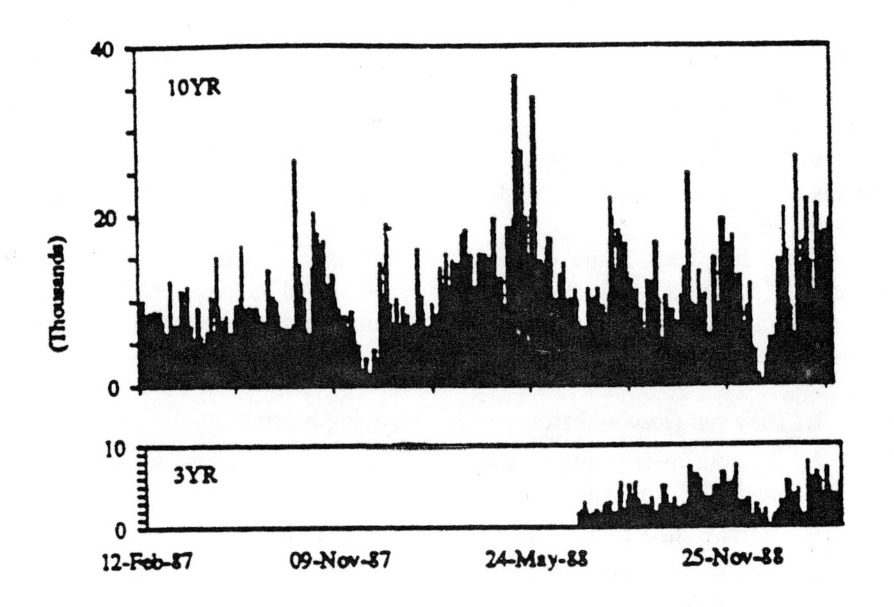

The SFE has recently introduced a deliverable futures contract on bonds issued by the semigovernment authorities. The pricing and delivery details (see Appendix C) of this contract differ substantially from those of deliverable contracts overseas, primarily as a result of the Australian market trading on yield, not price. The semigovernment contract started trading late October 1989 with contracts for March, June, September, and December expiry in 1990.

Managing a Portfolio's Risk and Return

We will now consider a portfolio that is out of balance with the index and use either futures or physicals to rebalance the portfolio. We will "test" the rebalanced portfolio's returns by comparing its response to various yield curve shifts over a holding period to the returns of the index. The index we will aim to track is the DBSM Bond Index (0+ years).

Initial Portfolio

Consider a portfolio with $100,000 face value. This portfolio's modified duration is less than that of the index (see Table 2) and thus will underperform the index if rates moved down. Figure 2 shows the percentage holding of the portfolio and the contribution to the total modified duration in the various term to maturity ranges. We note that the portfolio has a greater percentage of its securities with short term to maturities than the bond index. The returns over a two-month period for the portfolio and the index are given for a number of yield curve shifts in Table 3. The yield curve shifts are shown in Figure 3. We observe that the portfolio return is close to the index when yields do not change over the two months; the small difference in return is due to the 13 basis-point difference in the internal rate of return (see Table 2) providing an approximate 2 basis-point difference in return over 2 months.

However, the portfolio's returns differ from the index's when there are yield curve shifts over the two-month period. Since our portfolio is short modified duration by about 0.17 compared to the index, then with a 50 basis-point drop in yields we expect our portfolio to underperform the index by about 0.17

Table 2

	IRR	Moderate Duration	Convexity
Bond Index	13.67	3.40	0.200
Initial Portfolio	13.80	3.23	0.186

Figure 2 **Percentage of Portfolio and Bond Index in Term to Maturity Ranges**

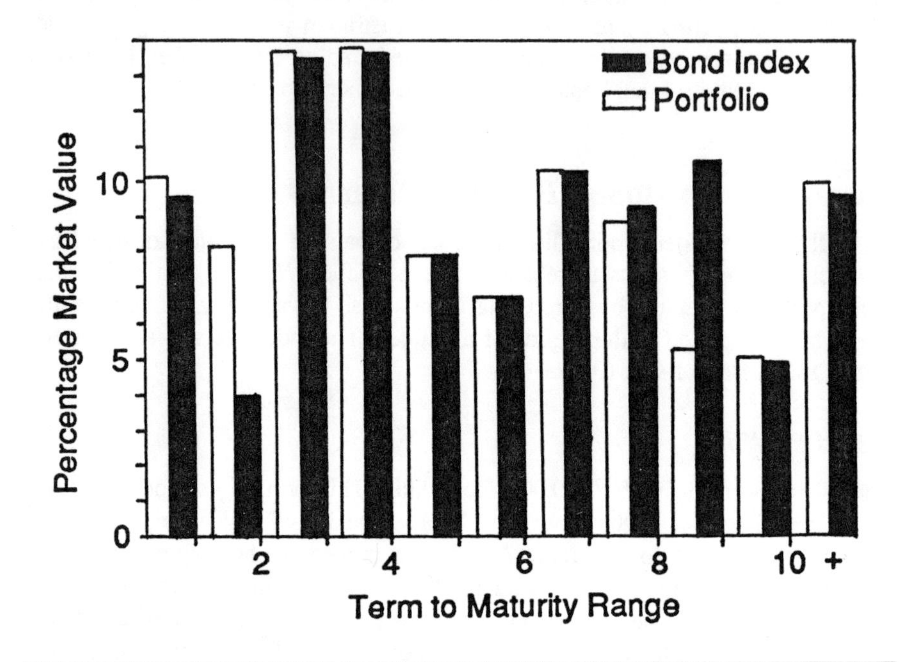

× 0.5 percent which is 8.5 basis points. When we add in the extra 2 basis points higher return due to the extra IRR, we expect our portfolio to underperform the index by 6.5 basis points in the two-month period. The data in Table 3 show that we actually underperform by 7 basis points. Since the portfolio returns will not track the returns from an index, we now investigate methods to restructure the portfolio's holdings with the aim of reducing the tracking error for various yield curve scenarios. The methods we will present to rebalance the portfolio can also be used to assist in managing the risk/return tradeoff when adopting an active benchmark management approach.

Method of Adjusting a Portfolio's Characteristics

The structural characteristics of a portfolio that we will use are the internal rate of return (IRR), modified duration, and convexity. The IRR provides a measure of the yield (or return) from the total portfolio over time with no movements in the term structure. Modified duration and convexity provide a measure of the

Figure 3 Yield Curve Scenarios

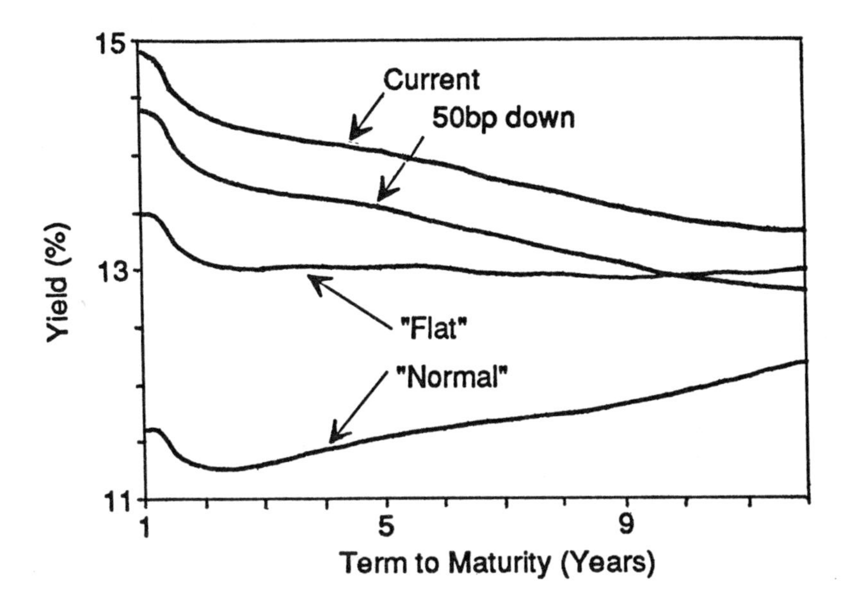

first and second order effects giving a change in market value with yield movements. In addition convexity also is related to a portfolio's response to twists in the yield curve. Since the Australian yield curve has experienced large movements in the long end (yields between 12 and 15 percent) and substantial whipping around in the short end (difference in the ten year and one year rates lie between –350 basis points and +350 basis points), we will focus our attention on the adjustment of a portfolio's modified duration and convexity. We note that the

Table 3

	No yield movement	50 basis points down	Flat yield curve	Normal yield curve
Bond Index	2.212	3.894	6.402	7.859
Initial Portfolio	2	–7	–15	–21

Note: Portfolio returns as basis points from bond index.

impact of futures on a portfolio's IRR should also be monitored when using futures, but we will leave the details for discussion in a forthcoming paper.

We note that the IRR can be approximated by the modified duration weighted sum of the yields

$$IRR = \frac{\sum N_i \times P_i\,(Y_i) \times MD_i \times Y_i}{\sum N_i \times P_i\,(Y_i) \times MD_i} \tag{1}$$

where the symbols are defined in Appendix B. A more accurate expression can be obtained if second order convexity effects are included in the approximation (see Appendix A).

The modified duration of a portfolio of bonds is given by the sum of the market value weighted modified durations of each security in the portfolio:

$$MD_{pfolio} = \sum N_i \times P_i\,(Y_i) \times MD_i / Value \tag{2}$$

Similarly, the portfolio's convexity is given by

$$Conv_{pfolio} = \sum N_i \times P_i\,(Y_i) \times Conv_i / Value \tag{3}$$

When we include futures in our portfolio, we add an extra term that relates to its contribution to the total portfolio's interest-rate sensitivity. Thus

$$MD_{pfolio} = \sum N_i \times P_i\,(Y_i) \times MD_i / Value + Value_{fut} \times MD_{fut} \times N_{fut} / Value \tag{4}$$

A similar expression is obtained for the convexity. The IRR is adjusted by the contribution to the portfolio return due to differences between the income from the bond that the futures contract is on and the cost of borrowing money to buy the bond (i.e., the cost of carry used to determine futures fair value), assuming the futures contract is priced at fair value. When the yield curve is inverse, the futures alone experience a loss over time if the yield of the underlying physicals and the repo rate remains constant, and the opposite applied when the yield curve is positive.

Using equation (4), we can calculate the number of futures contracts required to adjust the portfolio modified duration,

$$MD_{adjustment} = Value_{3YRfut} \times MD_{3YRfut} \times N_{3YRfut} / Value + Value_{10YRfut} \times MD_{10YRfut} \times N_{10YRfut} / Value \quad (5)$$

A similar expression can be obtained for the convexity adjustment by replacing the modified duration terms in equation (5) with convexity,

$$Conv_{adjustment} = Value_{3YRfut} \times MD_{3YRfut} \times N_{3YRfut} / Value + Value_{10YRfut} \times MD_{10YRfut} \times N_{10YRfut} / Value \quad (6)$$

We now have two equations with two unknowns, and these can be solved to find the number of 3- and 10-year futures contracts required.

Adjust Portfolio Duration

Index Tracking

We observed from Table 3 that our portfolio did not obtain the same return as the index when there are yield movements. We now attempt to adjust our portfolio holdings to that we have a closer match to the modified duration of the index.

Using futures we find that we require either 32 *or* 72, 10- or 3-year contracts respectively; the number of contracts have been rounded to the nearest integer. We add these futures to the portfolio and recalculate the portfolio's return (see Table 4). We observe that the portfolio with futures now obtains returns that are considerably closer to the index's when the yield curve shifts down by 50 basis points. This is due to the property of modified duration matching providing similar responses to small parallel yield curve shifts. We note that the addition of futures in an inverse yield curve environment results in a small drop of return if bond yields do not change over the time period. This is due to the fact that a fair priced futures will have a lower yield (higher price) than the physicals when purchased, and the futures closeout yield should converge to the physicals at closeout.

If we chose to adjust our portfolio holdings using bonds, we could raise our modified duration so that it is closer to the index by selling some of the shorter-term securities and buying some longer-term securities. However, we have to consider the minimum face value we can trade with reasonable cost and liquidity. For this discussion, we will allow trades of physicals in parcels of $1,000 face value.

We, therefore, lengthen the portfolio to bring its portfolio's modified duration closer to the index's; by trading 5 percent of the portfolio, we can bring the modified duration to 3.42, which is just greater than the index's modified duration. The returns for this portfolio are also shown in Table 4. We note that the portfolio does not tract the index as accurately as the portfolio with futures for the parallel yield curve shift of 50 basis points down. This is primarily due to the restriction on the parcel size of trades on physicals compared to the contract size for futures; with the futures we can get a closer match of modified duration. We could, of course, trade a considerably greater portion of the portfolio to accurately match the index modified duration, but eventually the transaction costs outweigh the benefits to be achieved.

When adjusting the modified duration we have restructured our portfolio to enable it to have less tracking error for the parallel yield curve shift considered. However, the tracking error for nonparallel yield curve shifts is still high, as can be seen in Table 4.

Benchmark Management

If our view is that rates will generally move down in a parallel shift, then we could increase the portfolio's modified duration to be greater than the index's and thus gain enhanced price increases. However, we must be aware of the impact of underperformance of the benchmark index if our view is incorrect and rates move up.

Modified duration adjustment is often considered for views on parallel yield curve shifts. Alternatively, we could use convexity adjustments when taking views on parallel shifts to enhance returns with both raising and falling rates; this will be discussed in the next section.

Unlike index tracking management where we are looking at minor rebalancing of the portfolio every month or so, with an active strategy we may be taking short-term views for time periods down to hours. The use of futures for enhancing returns with short-term market movement forecasts will be attractive due to the higher liquidity and lower transaction costs.

Adjusting Portfolio Duration and Convexity

Convexity is the second order price sensitivity effect to yield changes; it accounts for most of the curvature in a graph of the change in price versus changes in yields. Convexity is also a measure of the dispersion of the time squared weighted cashflows of a security or portfolio, and it is through this property that it is used, in conjunction with modified duration adjustments, to tailor a portfolio's response to nonparallel yield curve shifts. Since we have two bond futures contracts to use, we can adjust both the modified duration and convexity.

Index Tracking

Our portfolio's modified duration and convexity differ from the index's by 0.17 and 0.0125 respectively. We calculate by solving equations (5) and (6) that we require 27 and 10, 10-year and 3-year contracts respectively, to bring our portfolio's modified duration and convexity in line with the index. The returns obtained for the yield curve shifts are given in Table 5. We now see that the returns for nonparallel yield curve shifts are now closer to the index than those obtained from a simpler duration adjustment (see Table 4).

We also calculated the trades of the physicals required to match the index modified duration and convexity. Because of the minimum parcel size of $1,000 used in this discussion, we could not obtain an exact match. An optimization technique known as integer programming, which can take into account the minimum transaction and parcel sizes and transaction costs was used to determine the portfolio that had the closest modified duration and convexity match to the index. In the application of this technique, we have also considered the structural aspects that lead to a close match of the IRR. We are required to trade 5 percent of the portfolio to match its characteristics to the index. The returns from this portfolio are given in Table 5, and we see that this portfolio also closely tracks the index for both parallel and nonparallel yield curve shifts.

Adjustment of the portfolio using physicals is constrained by the minimum parcel size for trades; this has a greater impact for small portfolios. A combination of adjustments using physicals to provide the major adjustments to the portfolio and then a small number of futures contracts to fine-tune the adjustments may be more desirable to managers who want only small exposures in futures; this will provide better results than when adjustment is done using physicals alone.

Table 4

	No yield movement	50 basis points down	Flat yield curve	Normal yield curve	Moderate duration	Convexity
Bond Index	2.212	3.894	6.402	7.859	3.40	0.200
Portfolio + 3YR futures	–1	–1	13	18	3.40	0.191
Portfolio + 10YR futures	0	0	–4	–6	3.40	0.206
Portfolio + physicals	1	2	–12	–17	3.42	0.210

Note: Portfolios return as basis points from bond index.

Table 5

	No yield movement	50 basis points down	Flat yield curve	Normal yield curve	Moderate duration	Convexity
Bond Index	2.212	3.894	6.402	7.859	2.40	0.200
P'folio + 3YR & 10YR futures	0	0	–1	–2	2.40	0.201
Optimised Portfolio	1	1	–4	–6	2.42	0.201

Note: Portfolios return as basis points from bond index.

Benchmark Management

If we are taking a view on nonparallel yield curve shifts, then we can rebalance the term to maturity structure of the portfolio to increase or decrease our exposure in various maturity ranges, depending on our view. For example, if we believed that long and short rates will remain constant and that medium-term rates will decrease, then we could increase our exposure in the medium maturity range and decrease our exposure in either or both of the short and long maturity ranges; the choice would depend on one's risk management style. When adjusting the maturity structure of the portfolio we should monitor the modified duration and convexity as an aid to risk quantification.

In the previous section we considered duration adjustments to add value from correct prediction of parallel yield shifts. Alternatively, we could leave our portfolio's modified duration at the index's value and adjust the convexity. If our portfolio convexity is greater than the index's, our portfolio will outperform the index for both parallel yield curve shifts up or down. However, we are exposed to underperformance of the benchmark for certain nonparallel yield curve shifts.

Generally a combination of both modified duration and convexity management would be considered under benchmark management.

What are the Tradeoffs?

With the ability to trade physicals over a wider term to maturity range, their use potentially provides a fine degree of portfolio restructuring. The availability of two bond futures contracts provides some limitations of their ability to fine-tune the term to maturity structure of a portfolio. One of the tradeoffs is the small contract size, the futures having a greater degree of fine-tuning in this regard.

The futures have high liquidity, and trades can be executed very quickly if one requires quick exposure for timely market movements. Additionally, the futures do not always trade at fair value. The slight mispricing can also be to one's advantage. However, the mispricing may go against us, and this must be considered.

The Squeeze

The squeeze at expiry of a bond futures contract occurs when, for example, a large long position in futures is held and then pressure is applied on the physical market to decrease the yield of the pricing bonds for the futures closeout. The pressure on the bond market is applied by initiating a large buy order in the physical market on the morning of the closeout, thus supply and demand will force the price up. The size of the long futures position held will be considerably greater than the number of purchased physicals required to move the market, thus the interest-rate risk taken on with the long physicals is more than offset by the greater holdings in the futures. This leads to gains for the long futures position and a profit or loss for the physicals purchased during the process, resulting in an expected net profit due to the relatively large futures position; the physicals may display a loss if the bond yields move down before the position can be sold, but the net profit should still be positive. There are some interesting variations of setting up and unwinding this play that can further enhance the players profit.

If we expect that the market may be distorted by such squeezes, then we can take advantage by either taking the correct side of the squeeze without actively being involved in its mechanics or reducing our exposure to that section of the market. For example, if you expect the market to pressure the closeout yield down, then a net long futures position will be profitable. Of course, if everybody did this then the price of the futures contract would decrease, leading to arbitrage pressure bringing the futures and physicals into fair value ranges.

In addition, some interesting switches between bonds may be available. For example, if you expect a squeeze will force the prices of the bonds in the futures pricing basket upwards, then a switch from slightly shorter bonds in the 8- to 9-year range into the 10-year bonds may add some value to a portfolio.

With the change in the pricing of the futures closeout (from the market perception of a fictitious 10-year bond) to the average yield of a basket of five bonds, such squeezes will be considerably more difficult to execute and have not occurred so far. The expected deliverable contracts should further increase the difficulty in executing squeezes as the number of deliverable bonds will probably be greater than five, and pressure has to be brought on all the bonds to force the price up. Also, as one is delivered physicals if a net long position is held at expiry, then the full interest-rate risk will be transferred to the long futures holder making the squeeze considerably harder to implement.

Basket Trading

Given that the current bond futures contracts are trading off the average yield of a basket of bonds, there are opportunities for setting up trading of the pricing basket to assist in arbitrage of the futures. We will now consider the characteristics of the basket and compare them to the futures. Taking the closing yields on 23/2/89, we have calculated the modified duration and convexity for both the basket and the futures for the 3- and 10-year contracts (see Table 6).

We note that an equally weighted 10-year basket is slightly shorter than the futures contract, and the opposite is the case for the 3-year basket. This leads to the requirement to manage the risk of different responses of the basket and the futures to yield curve movements if one is buying or selling the basket for the

Table 6

	Modified Duration	Convexity
3-Year Futures	2.41	0.074
3-Year Basket	2.51	0.084
10-Year Futures	5.47	0.433
10-Year Basket	5.11	0.389

purpose of arbitrage or locking into an implied reinvestment rate. For example, if one purchased $1,000 face value of each of the bonds in the 10-year basket, its response to instantaneous yield curve shifts would differ from the response of the 50 10-year futures. The basic techniques presented here for fund managers can be developed for the use of arbitrageurs in controlling these risks.

Another source of risk in buying and selling the basket is the timing of the purchase/sale of the bonds. If we cannot trade the issues in a timely fashion, then we are at risk of being exposed to market movements affecting the purpose of the trade. Thus the liquidity of the bonds in the basket should be addressed. All the bonds in the 10-year basket are reasonably liquid, but the Feb and July 92s are not as liquid as the rest in the 3-year basket.

Summary and Conclusions

In this chapter we have provided some methods for using physicals and futures to adjust a portfolio's holdings to meet particular management objectives. In particular, we have focused our attention for illustrative purposes on the index tracking management approach. We took a portfolio that was not in line with an index, then rebalanced it using duration and then duration and convexity considerations. We observed that the portfolio could be rebalanced to meet required objectives and that the use of futures had some additional advantage due to the smaller contract size, but the disadvantage is the limited range of term to maturities of the available contracts.

The application of the techniques was also discussed when considering a benchmark management approach. The range of possibilities for tilting the portfolio is a very extensive subject and only briefly touched. The one important feature is the management of the risk of underperformance of the benchmark index, and this can be controlled with the use of futures.[2]

[2] See footnote 1.

Some of the interesting and controversial topics with regard to the squeeze at expiry and the moves by the SFE to rectify the problems were discussed. We concur with the moves made by the SFE, first with the futures trading off a basket of bonds and, second, with the implementation of the deliverable contracts. The basket approach still has some problems with the different risk exposures between the basket and the futures. The implementation of the deliverable contracts will certainly provide some new considerations for fixed-interest portfolio management.

Appendix A: Internal Rate of Return

The IRR is the yield Y_{IRR} such that the total of the discounted cashflows, using IRR, is the current market value of the portfolio;

$$\sum N_i \times P_i\,(Y_{IRR}) - Value$$

$$- \sum N_i \times P_i\,(Y_i) \qquad \text{(A1)}$$

The exact solution of equation A1 requires an interactive numerical method that is best performed by a computer. A simple approximation can be made by noting that the difference in $P_i(Y_{IRR})$ and $P_i(Y_i)$ can be approximated by the price sensitivity to interest rate changes due to modified duration,

$$P_i\,(Y_{IRR}) - P_i\,(Y_i) = -MD_i \times P_i\,(Y_i) \times (Y_{IRR} - Y_i) \qquad \text{(A2)}$$

Using equations (A1) and (A2) we can rearrange to obtain

$$Y_{IRR} = \frac{\sum MD_i \times P_i\,(Y_i) \times N_i \times Y_i}{\sum MD_i \times P_i\,(Y_i) \times N_i} \qquad \text{(A3)}$$

Thus, the IRR is approximated by the modified duration value weighted yields.

A better approximation can be obtained by including the convexity term into equation (A2) giving

$$P_i\,(Y_{IRR}) - P_i\,(Y_i) = -MD_i \times P_i\,(Y_i) \times (Y_{IRR} - Y_i)$$

$$+\, Conv_i \times P_i\,(Y_i) \times (Y_{IRR} - Y_i)^2/2 \qquad \text{(A4)}$$

Substituting equation (A4) into (A1), we obtain a quadratic to solve for Y_{IRR}, giving

$$IRR = (B - \sqrt{(B^2 - 4AC)})/2A \qquad \text{(A5)}$$

Where

$$A = \sum N_i \times P_i\,(Y_i) \times Conv_i/2$$

$$B = \sum N_i \times P_i\,(Y_i) \times (MD_i \times Conv_i \times Y_i)$$

$$C = \sum N_i \times P_i\,(Y_i) \times (MD_i \times Y_i + Conv_i \times Y_i{}^2/2)$$

Appendix B: Nomenclature

Y_i — yield of i th security.

$P_i(Y_i)$ — gross (not capital) price of the i th security at yield Y_i.

N_i — number of $100 units of face value.

MD_i — modified duration of i th security.

$Conv_i$ — convexity of the i th security.

$Value$ — market value of portfolio = $\sum N_i \times P_i\,(Y_i)$.

$Value_{fut}$ — value of holdings in a futures contract.

Appendix C: Details of the Semigovernment Contract

Delivery: A net holder of short futures at closeout will be required to deliver a semi(s) from a list of eligible issues. This list will be produced by the SFE 4 months before the closeout month of a contract and will be based on the selection critera listed below.

Deliverable Issues:

- semi matures in 1995.
- stock must be inscribed.
- stock must be standard bullet loan inscribed stock.
- stock must pay equal coupons on semiannual basis.
- stock must have explicit State or Federal Government Guarantee (except Telecom).
- the guarantee must extend to the stock's maturity.
- the stock must be capable of being marked in Sydney & Melbourne.
- minimum amount outstanding of $200,000.
- the stock must be issued by one of the following authorities:
- Telecom.
- MMBW.
- Govt. of NT.
- NSWTC.
- QTC.
- SAFA.
- SECV.
- TASCORP.
- VICFIN.
- WATC.

Delivery Period:

A short futures holder must inform the Clearing House (ICCH) by 4:00 PM of the closeout day the semis he will deliver. The ICCH will inform the long holders which semis they will receive by 12:00 noon the next day. Settlement of delivered issues will be 5 business days after closeout (i.e., there is only one delivery date).

Margining:

Margining will be based on the change in the price of a hypothetical 12 percent coupon bond valued with n coupon payments. It is believed that the initial value of n will be 10 for the first 4 contracts listed.

Delivery Invoice Price:	The price a long futures holder will pay for delivered issues will be determined by the standard Reserve Bank of Australia formula for the delivered bond calculated to the settlement date using the closeout yield of the futures (not the market yield).

CHAPTER 10

The Performance of Currency-Hedged Foreign Bonds*

Lee R. Thomas III**
Investcorp Bank, E.C.

Investors who hold fixed-income securities denominated in foreign currencies incur the usual bond risks—primarily credit risk and interest rate risk—plus the risk that the foreign currency in question will depreciate against their base currency. Because of this exchange rate risk, it may seem difficult to take advantage of opportunities in foreign markets. A U.S.-dollar investor, for example, may believe that particular foreign bonds offer unusual value, or may simply wish to reduce exposure to changes in domestic interest rates by diversifying into foreign bonds. Yet he may hesitate to invest abroad if he thinks foreign currencies are likely to depreciate against the U.S. dollar.

Fortunately, portfolio managers need not forgo the opportunities offered by foreign bonds just because of exchange rate risk. They can instead use foreign bonds that have been *synthetically redenominated* into U.S. dollars. A synthetic dollar bond is exposed to changes in foreign interest rates, but it is relatively insensitive to changes in domestic interest rates. Consequently, such a bond of-

* This chapter is an adaptation of Lee R. Thomas III, "The Performance of Currency-Hedged Foreign Bonds," *Financial Analysts Journal*, May/June 1989. Copyright 1989 by the Financial Analysts Federation.

** Lee Thomas is a member of the management committee, responsible for trading and arbitrage at Investcorp Bank E.C. When this article was written, he was an Executive Director, Financial Strategies Group, at Goldman Sachs International Limited.

fers an attractive diversification outlet plus an opportunity to earn capital gains if foreign interest rates decline. At the same time, the synthetic dollar bond, by design, will have little exchange rate risk.

A synthetic dollar bond can be constructed by combining short-dated forward exchange contracts with longer-dated foreign currency bonds.[1] This technique for reducing or eliminating exchange rate risk is called *a rolling forward hedge*. Briefly, the forward contract creates a foreign currency liability equal in value to the foreign currency asset—the bond. But because its duration is short, the forward obligation does not substantially alter the bond's interest rate exposure.

This chapter examines the performance of synthetic dollar bonds—foreign bonds hedged with rolling forward contracts. In all cases, the perspective is that of a U.S.-dollar investor holding 10-year government bonds.[2] We show that the arguments for including hedged foreign bonds in an investment portfolio are compelling. Hedged foreign bonds have offered dollar investors about the same return, with far less risk, than unhedged foreign bonds.

Risk and Return

Imagine that a dollar-based investor buys a nondollar bond at the beginning of the month. She hedges her exchange rate risk by selling the bond's current foreign currency value, plus future coupon payments or accrued interest, for one-month delivery against U.S. dollars. The forward exchange rate for this hedging transaction will reflect the foreign currency's discount or premium. At month end, she sells the foreign bond and delivers the foreign currency realized from its sale to satisfy her forward exchange contract.

We can divide the total return on this hedged bond transaction into four parts:

1. interest income earned, including accrued but unpaid interest;
2. the capital gain or loss on the bond;
3. the discount or premium on the foreign currency sold forward; and
4. the foreign exchange rate change during the month, applied to the capital gain or loss on the bond.

[1] The mechanics of currency-hedging foreign bond investments are described in L. Thomas, "International Bonds: Stripping Away Current Risk," *Investment Management Review*, March/April 1988.

[2] For complementary analyses, see L. Thomas, "Currency Risks in International Equity Portfolios," *Financial Analysts Journal*, March/April 1988 or Thomas, "The Performance of Currency Hedged Foreign Equities" (Goldman, Sachs & Co., New York, July 1988). For currency hedging from a U.K. sterling or Japanese yen perspective, see Thomas, "The Role of Currency Hedging in U.K. Investment Portfolios," in *Competitive Strategies for Asset/Liability Management* (London: IFR Publishing Ltd., 1988) and Thomas, "Rules for Global Asset Allocation," *Security Analysts Journal* (Security Analysts Association of Japan), forthcoming.

Item (2), the capital gain or loss, will depend on the bond's duration and on the change in foreign interest rates during the month. The sum of the capital gain or loss and Item (1)—interest income earned—will equal the return to the bond measured in its local currency.

Item (3), the discount or premium on the foreign currency, is roughly equal to the one-month interest rate differential (on Euro-deposits) between the U.S. dollar and the currency in which the bond is denominated. If the foreign one-month rate is lower than the U.S. dollar rate, then the foreign currency will sell at a forward premium. This increases total return, measured in dollars. If the one-month foreign interest rate exceeds the U.S. rate, then the foreign currency will sell at a discount, reducing dollar return.

Item (4) represents the investment's foreign exchange rate risk. Because the currency hedge, accomplished by selling foreign exchange forward for dollars, covers only the initial value of the bond and one month's interest income—not the increase or decrease in the bond's market value—the investment is exposed to exchange rate changes on the capital gain or loss. But this element will ordinarily be small compared to the other three.[3] We can safely neglect it in the discussion that follows.

The total variability of a hedged foreign bond's return thus equals the variability of the bond's return measured in its local currency (interest plus capital gain or loss), plus the variability of the forward exchange discount or premium, plus a component contributed by the covariation between these components. In practice, the risk of a hedged foreign bond is due almost entirely to variation in local-currency return; that is, the risk is largely foreign interest rate risk. The contribution of changing forward foreign exchange discounts or premiums is negligible.

Of course, most investors do not ordinarily hedge the exchange rate exposures embedded in their foreign bonds. For an unhedged investment, the total rate of return on the bond position will be the sum of the rate of return in local currency and the rate of change in the value of the foreign currency (measured in U.S. dollars).

Just how significant is the latter component? In practice, is exchange rate risk important to an international fixed-income investor? Or are the effects of exchange rate changes swamped by the effects of changing foreign bond prices?

Table 1 assigns the risk of 10-year unhedged foreign bonds to three categories: (1) variation in the bond's local currency return, primarily because of monthly capital gains or losses resulting from interest rate changes; (2) variation in the exchange rate; and (3) covariation. The uncertainty in unhedged foreign bond returns is, on average, more attributable to exchange rate risk than to foreign interest rate risk. In the case of the two most important nondollar markets—Japan and Germany—exchange rate risk is much more serious. Moreover,

[3] Moreover, you can reduce this by rebalancing the hedge frequently.

Table 1 Sources of Risk in Foreign Bonds: 1975–88*

Country	Local Currency Variation (%)	Exchange Rate Variation (%)	Covariation (%)
Germany	17	58	25
Japan	21	49	25
U.K.	38	39	22
France	17	72	11
Canada	65	18	17
Netherlands	22	68	11
Average	30	51	19

* Here risk is measured by variance of return rather than the standard deviation of return, as elsewhere in this chapter. We chose to use variance here to ensure that the sum of the risks contributed by interest rate changes, exchange rate changes and covariation equaled 100 percent of the total variation. All bonds in the analysis are approximately 10-year maturities.

exchange rate risk would be of even greater significance, in relative terms, to the returns on shorter-maturity bonds.

Individual Bond Performance

How have foreign bonds, hedged and unhedged, performed during the floating exchange rate period? Table 2 summarizes the data on foreign government bond risk and return from the beginning of 1975 through the end of 1988.

The unhedged foreign government bonds examined generally earned considerably more than the comparable U.S. Treasury bonds. The average foreign 10-year bond returned 11.3 percent per year, or 240 basis points more than the return from 10-year U.S. Treasuries. Moreover, five of six foreign bonds returned more than U.S. Treasuries did. Only one foreign market—Canada—earned less.

Currency hedging is not designed to *enhance* the returns to a foreign bond position, but to *stabilize* them. There are no strong reasons to believe that hedging increases or decreases return in the long run, beyond the additional transaction costs it imposes.[4] Thus, it is not surprising to find no clear relation between hedged and unhedged returns. Our results show that hedged returns exceeded their unhedged counterparts in three cases, fell short in two cases, and were identical in one case. In five of the six cases, however, returns to the hedged foreign bonds exceeded returns to U.S. Treasuries; the average return of 11.2 percent exceeded the return on U.S. Treasuries by 230 basis points.

[4] Transactions costs are typically small in the foreign exchange markets.

The major advantage of currency hedging is reduced risk. Table 2 shows that the reduction in volatility (measured by the standard deviation of return) was substantial for all except Canadian bonds, and even these bonds enjoyed a moderate reduction in risk. The annualized return volatility of the German bonds, for example, dropped to 6.4 percent (hedged) from 15.7 percent (unhedged), and Japanese bond volatility went to 7.7 from 17.0 percent. In fact, the risk reduction afforded by hedging was large enough to reverse the riskiness of unhedged foreign bonds vis-a-vis U.S. bonds. Held unhedged, all the foreign bonds were riskier than U.S. Treasuries, often by a substantial margin. When hedged, five of the six foreign countries' bonds were less risky than U.S. Treasuries.

We can construct a summary measure of the relative attractiveness of an asset by dividing its average return by the standard deviation of its return. Although this ratio ignores some important questions—such as how much of the asset's risk remains after diversifying—it gives a rough measure of how much return a bond holder enjoyed per unit of risk borne. In the case of every foreign bond, hedging increased this return/risk ratio. Five of six hedged bonds outperformed U.S. Treasuries substantially; the return/risk ratio of the sixth (Canada) was only marginally worse. The average foreign bond's return/risk ratio roughly doubled to 1.49 (hedged) from 0.75 (unhedged).

Bond Portfolio Performance

Most dollar investors in foreign fixed-income securities hold bond portfolios rather than single issues. These investors may be less interested in a particular foreign bond's performance than in how a representative foreign bond portfolio would have performed relative to U.S. Treasuries. To find out, we examined a broadly diversified portfolio of the bonds shown in Table 2 over the 1975–88 period.[5]

Table 3 gives the results. When held unhedged, the foreign portfolio had a mean return of 12.1 percent, with a standard deviation of 12.4 percent. This produced a return/risk ratio of 0.98, a little better than that of U.S. Treasuries (0.94).

As Table 4 shows, hedging reduced the return of the foreign bond portfolio slightly, to 11.5 percent. But hedging cut the foreign portfolio's risk by more than one-half, to 5.3 percent. Consequently, the hedged foreign bonds registered a return/risk ratio of 2.19. This result is better than twice the performance ratio recorded by an individual unhedged foreign bond, by the unhedged foreign portfolio, or by U.S. Treasuries.

[5] The foreign bond portfolio weights are: Japan, 30 per cent; Germany, 25 per cent; France and the U.K., 15 per cent each; Canada, 10 per cent; Netherlands, 5 per cent.

Table 2 10-Year Government Bond Performance, 1975–88 (U.S. Dollar Investor's Perspective)

Country	Unhedged Mean Return (%)	Hedged Mean Return (%)	Unhedged Risk (%)*	Hedged Risk (%)	Unhedged Return Risk	Hedged Return Risk
U.S.	8.9	–	9.5	–	0.94	–
Germany	10.6	12.0	15.7	6.4	0.67	1.87
Japan	15.7	12.6	17.0	7.7	0.92	1.64
U.K.	12.9	12.5	18.3	11.3	0.71	1.10
France	9.6	9.6	13.7	5.8	0.71	1.66
Canada	8.1	8.2	11.6	9.3	0.70	0.88
Netherlands	11.1	12.0	14.6	6.8	0.76	1.78
Average	11.3	11.2	15.2	7.9	0.75	1.49

Table 3 Unhedged Foreign Bond Performance in Subperiods

	Foreign Portfolio			U.S. Treasuries		
Subperiod	Mean Return (%)	Risk (%)	Return Risk	Mean Return (%)	Risk (%)	Return Risk
1975–79	13.2	9.8	1.36	5.7	5.4	1.06
1980–84	2.5	12.3	0.21	10.2	12.2	0.83
1985–88	22.7	14.5	1.56	8.9	9.5	0.94
Entire Period	12.1	12.4	0.98	8.9	9.5	0.94

Table 4 Hedged Foreign Bond Performance in Subperiods

	Foreign Portfolio			U.S. Treasuries		
Subperiod	Mean Return (%)	Risk (%)	Return Risk	Mean Return (%)	Risk (%)	Return Risk
1975–79	10.5	3.9	2.66	5.7	5.4	1.06
1980–84	13.7	5.4	2.54	10.2	12.2	0.83
1985–88	10.0	6.3	1.58	11.3	9.5	0.94
Entire Period	11.5	5.3	2.19	8.9	9.5	0.94

These results strongly suggest that hedged foreign bonds have been a better long-run investment than unhedged foreign bonds.[6] To understand the significance of the risk-return tradeoff from hedged versus unhedged diversification, consider a dollar investor who commits 20 percent of his resources to the foreign bond portfolio and the remaining 80 percent to U.S. Treasuries, then hedges some of his foreign bonds' currency risks. Figure 1 shows what his mixed dollar/foreign bond portfolio's risk would have been, depending on the share of the foreign bonds he hedged.

Note that investing 20 percent of the portfolio in foreign bonds lowers the volatility by 0.6 percentage points, even without currency hedging. The volatility of the U.S. Treasury portfolio is 9.5 percent, compared to 8.9 percent for the diversified unhedged portfolio. Hedging consistently reduces the riskiness of the diversified portfolio; the more hedged the portfolio, the greater the risk reduction. Combining diversification with full currency hedging lowers the volatility of the fixed-income portfolio by 1.3 percentage points, or roughly twice as much as diversification alone. An *unhedged* investor who committed 20 percent of his resources to foreign markets would have enjoyed only *half* the risk reduction he could have had if he had diversified *and* currency-hedged.

Hedged Bonds in Internationally Diversified Portfolios

Most investors are familiar with the idea of reducing risk by diversifying. Many fixed income investors spread their credit risks among corporate, agency, and government issues. Investors can also reduce interest rate risk to some extent by investing in different segments of the domestic yield curve. Unfortunately, the opportunity to diversify in this way within the U.S. bond market is limited, because all interest rates often move in the same direction when the yield curve shifts.

One solution to the problem is to hold some assets in foreign bonds. Foreign bonds carry their own interest rate risks, but they are not directly exposed to U.S. interest rate changes. As long as foreign and U.S. interest rates are not perfectly correlated, some of the interest rate risk will be self-canceling. In general, correlations between the returns on debt instruments in different national markets are much lower than the return correlations between instruments within the U.S. market.

Unfortunately, substituting foreign interest rate risk for dollar interest rate risk by adding unhedged foreign bonds to a dollar portfolio also adds exchange rate risk to the portfolio. Because U.S.-dollar exchange rate changes are highly

[6] An investor who correctly predicted when foreign currencies would appreciate and used this information to hedge selectively would of course have done much better. There is some evidence that outforecasting the forward foreign exchange market is possible. See, for example, J. Bilson and D. Hsieh, "The Profitability of Currency Speculation," *International Journal of Forecasting* 3 (1987), No. 1.

Figure 1 Risk Reduction from Diversification, 1975–88

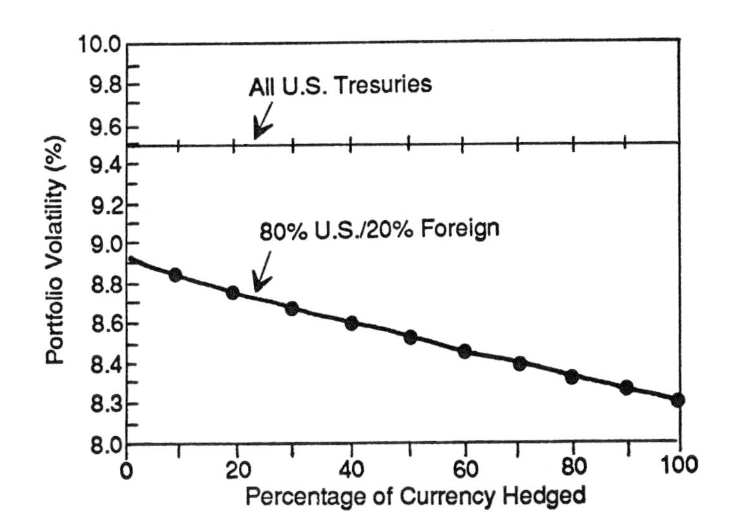

correlated, exchange rate risk cannot be eliminated by spreading foreign bond holdings across many countries.[7] The gains from unhedged international fixed-income diversification can thus be quickly swamped by exchange rate risk.

Hedged diversification, however, is another story. Hedged foreign bonds offer an opportunity to diversify a fixed-income portfolio's interest rate risk without adding exchange rate risk. And the potential for risk reduction is substantial. With the exception of Canadian-dollar bonds, the return correlation between 10-year U.S. Treasuries and each of the foreign bonds in our study is below 0.50.[8] Moreover, as Table 2 showed, hedged foreign bonds are usually less volatile than similar-maturity U.S. Treasury bonds. Together, these characteristics suggest that hedged foreign bonds offer an attractive opportunity for U.S. investors who wish to reduce interest rate risk in their portfolios.

[7] Currency-hedging reduced the volatility of the foreign bond portfolio by about as much as it reduced the volatility of investments in the average single foreign bond market. This indicates that exchange rate risks are systematic within the foreign portfolio. That is, exchange rate risks cannot be eliminated merely by spreading your foreign bond holdings among many countries.

[8] The hedged foreign bonds' correlations with U.S. Treasuries from 1975 through 1988 are: Germany, 0.42; Japan, 0.25; U.K., 0.19; France, 0.22; Canada, 0.69; and Netherlands, 0.26. In recent years the correlations have increased. This is a predictable side effect of informally managed exchange rates ("target zones").

Figure 2 Unhedged Diversified Portfolios, 1975–88

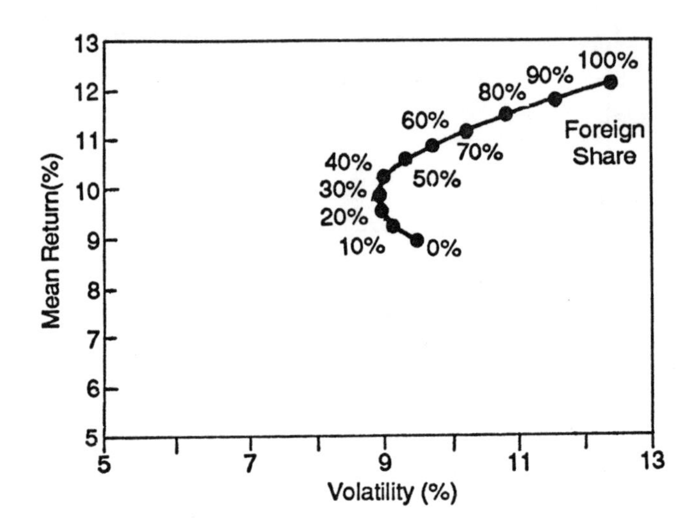

Hedged and Unhedged Diversification

Figure 2 illustrates how mixed portfolios containing U.S. Treasuries and unhedged foreign bonds have performed. It shows the average return and volatility of a portfolio that combines a sample foreign bond portfolio with a 10-year U.S. Treasury. The share allocated to foreign bonds varies from 100 percent (all foreign bonds) to zero (all U.S. Treasuries) in 10 percent increments.

The results are broadly consistent with previous studies of international investing, which have argued that foreign diversification can increase return at a given level of risk. Because the unhedged foreign bond portfolio had a higher average return than Treasuries—by 320 basis points per year—the greater the investment in foreign bonds, the higher the portfolio's return was.[9] Investing about 30 percent of assets abroad during 1975–88 would have lowered bond portfolio risk modestly and simultaneously increased return. Any diversified portfolio with less than 55 percent in foreign bonds would have realized both higher returns and lower risk than U.S. Treasuries alone.

The least risky portfolio—70 percent dollar and 30 percent nondollar bonds—had an annualized volatility unusual market conditions during one sub-

[9] Of course, historical returns are notoriously poor predictors of future returns. You should not conclude that the average foreign bond will have a substantially higher return in the future than a comparable maturity U.S. Treasury.

period. Rather, the hedged portfolio performed consistently better than either its unhedged counterpart or U.S. Treasuries. All in all, it appears that, by hedging, a U.S. portfolio manager who diversified into foreign bonds outperformed undiversified portfolio managers in each of the short-run periods examined, as well as in the long run.

Implications for Global Bond Investors

Our results are pertinent to selecting a strategy for global asset allocation. In particular, they stress the fact that participation in foreign fixed-income markets does not have to entail exchange rate risks. Forward exchange contracts can be used to currency-hedge nondollar bonds.

It is advisable to select foreign bonds independently of exchange rate expectations. The currency exposure of the portfolio can then be considered. A decision not to hedge is, in effect, a decision to add an active foreign exchange position—a currency overlay—onto the underlying hedged foreign bond portfolio. Whether or not this is advisable depends on the perceived risks and prospective rewards of bearing currency exposures.

Essentially, we suggest that an unhedged foreign bond position be thought of as a combination of a hedged bond investment and a separate long foreign currency position; forward contracts are the tool used to unbundle these two sources of risk and return. Considered in this framework, our main conclusion is that hedged foreign bonds have been a good buy-and-hold investment, but foreign currency has not. A foreign bond investor who never currency-hedged suffered more volatility, with no significant return advantage, than an investor who routinely hedged.

One strong implication of our result is that foreign currencies do not naturally enter into an investor's core holdings for the long run; they are a tactical—not a strategic—asset class. This means that investors who have no opinion concerning the future path of exchange rates should avoid unhedged foreign currency positions.[10] Exchange rate exposures add considerable risk to a bond portfolio—even if they are spread over many currencies. To justify bearing them, an investor *must* expect foreign currencies to outperform their forward rates. And, of course, foreign currencies cannot be expected to outperform their forwards consistently. As a result, currency-hedging foreign bond holdings should always be the base case.

This does not of course mean that an investor should never assume exchange rate risk. An investor who expects particular currencies to outperform

[10] Contrast this with, for example, equity investing. An investor who does not have strong opinions concerning which stocks are likely to outperform the market should diversify over many issues. But an investor who has no currency views should not diversify. Instead, he should denominate all of his portfolio—synthetically, where necessary—in his domestic currency.

their forwards at particular times can (1) decline to hedge, (2) hedge only partially, (3) over-hedge or (4) cross-hedge, depending on the strength of her exchange rate views.[11] But a decision to assume an exchange rate exposure should be a measured one, which weighs exchange rate risks against the prospective rewards of an open currency position. Exchange rate risks should never be borne by default, just because one happens to own foreign bonds.

[11] By cross hedging, we mean transforming one risky exchange rate exposure into another, more attractive exposure. For example, a dollar investor who wanted to hold German bonds but expected the pound to appreciate could buy a Bundesanleihen and simultaneously sell marks forward for sterling. The result would be a synthetic bond with German interest rate risk but pound/dollar rather than mark/dollar exchange rate exposure.

SECTION II

PRACTICAL CONSIDERATIONS

CHAPTER 11

International Benchmarks*

John Meier
BARRA

Introduction

Two of the major issues facing international investment are evaluating investment management performance and identifying portfolio risks. The choice of a benchmark affects both of these issues directly. This chapter looks first at the uses of international benchmarks from a traditional portfolio manager's perspective, and examines how a benchmark affects an international manager's investment behavior. We then look at some recent developments in the field of international investing that have influenced the ways international managers view benchmarks. We also examine how the increased use of quantitative techniques in international investing has affected the use and construction of international benchmarks.

Current Benchmarks

Before we get into these topics, let's look at some of the benchmarks now used to evaluate the performance of international managers. First, remember that benchmarks are just portfolios—a list of assets and investment weights. What differentiates a benchmark portfolio from other portfolios is the way it is used.

* Portions of this material originally appeared in an article published in *Pensions & Investments*, April 30, 1990.

The three most popular benchmarks now are the Morgan Stanley Capital International (MSCI) Indices, the Financial Times/Goldman Sachs (FT) Indices, and the Salomon Brothers/Frank Russell Indices.

MSCI Indices cover nearly 1,500 assets and 22 countries including 21 individual country indices; while Singapore and Malaysia are grouped as one local index. The sum of these assets composes the World Index, and each asset is weighted according to its market capitalization. MSCI also maintains several other indices that are country subsets of the World Index. The MSCI benchmark most often used for the international investing of U.S. funds is the Europe, Australia, and Far East Index (EAFE). This index consists of all World Index names not in the U.S., Canada, or South Africa, totalling around 1,000. Again, each asset in the EAFE is weighted according to its market capitalization.

The FT indices cover 24 countries and about 2,500 assets. The sum of the assets composes the FT World Index. However, unlike the MSCI Indices, the assets are not weighted according to their market capitalization. The capitalizations of certain assets are adjusted to reflect elements like foreign ownership restrictions and cross holdings. Like MSCI, FT maintains indices that are subsets of the World Index. The FT benchmark used for international investing of U.S. funds is the Europe and Pacific (EURPAC). This index consists of all the assets in the FT World Index not in the U.S., Canada, Mexico, or South Africa, totalling some 1,700 assets. As with EAFE, each asset is weighted according to its FT-adjusted capitalization.

The SAL/FR Indices cover 23 countries. Unlike the MSCI and FT indices, SAL/FR actually has *three* world indices, the PMI (Primary Market Index), the EMI (Extended Market Index), and the BMI (Broad Market Index). As of March 31, 1990, the same countries are in each index, but the assets included differ as follows: The PMI consists of about 1,250 larger capitalization, more liquid stocks. The EMI consists of about 2,900 smaller assets and thus aims at being an international small capitalization index. The BMI is the sum of the PMI and the EMI. As with the FT indices, the weights of the assets in the SAL/FR indices are adjusted by market capitalization. These adjustments include liquidity, cross holdings, and ownership restrictions. Other variations of these indices are used, and several organizations have developed indices that are not as widely used as those mentioned above.

Traditional View of International Benchmarks

In this section, we discuss the foundations of international managers' traditional view of benchmarks. How have most international managers viewed their benchmarks? The answer is—not frequently. Until fairly recently most international managers paid little attention to the benchmark they were supposed to be following. No matter what benchmark was used, the manager's investing behavior would not change. In fact, we have seen cases where a sponsor changed the benchmark to one dramatically different from the previous one, and the manager's portfolio or investment behavior did not change significantly. Many managers still see benchmarks as useless and exhibit associated behavior. What explanations could account for this type of behavior?

We can identify three factors. First, most managers see their job as managing funds to maximize total return, *not* to provide active return relative to a benchmark. Obviously, a goal of maximizing total return results in indifference to the benchmark and leads to the behavior mentioned above.

We consider this approach inappropriate. The sponsor has set a benchmark and seeks a portfolio with a risk-and-return profile similar to that of the benchmark. The manager's portfolio could have dramatically different characteristics, and these could alter the risk of the sponsor's overall fund or portfolio.

This situation, however, is not entirely the manager's fault. The sponsor has not taken an active role in controlling the behavior of his or her international managers and, thus, the risk profiles of their portfolios. Indeed, investment behavior and risk profiles relative to benchmarks that would be unacceptable if provided to the sponsor from a domestic manager are often readily accepted if they come from an international manager. Inconsistencies like these by the sponsor contribute to manager behavior that works against the goals of the fund. If sponsors showed more concern about differences between the benchmark they have chosen and the manager's portfolio, the manager would be more interested in the benchmark and the risk-and-return profile of the managed portfolio relative to the benchmark.

Finally, a manager's apparent disregard for the benchmark could be due to its composition and how it could be viewed as a business risk, or the risk of having funds taken away from the manager. For example, the EAFE Index has a weight in Japan of over 50 percent. Many European and North American investors believe that the Japanese equity market is significantly overvalued and will eventually decline significantly. If the Japanese market crashes and the manager is 50 percent invested in Japan following EAFE, the manager could have a good active return by outperforming EAFE but wind up with a terrible total return. It is very possible that the sponsor would take the funds away from this manager because of poor total return, not because of poor performance.

To avoid this calamity, the manager may structure portfolios to outperform EAFE if the Japanese market does poorly and underperform EAFE if the Japanese market does well. In the event of a disaster in Japan, the manager will most likely still have a positive total return and retain the funds under management. Thus, a manager's portfolio may take on significant active risk relative to the benchmark in order to reduce the manager's business risk.

Recent Developments

Increased Interest in International Investments

The traditional view of international benchmarks discussed above has been altered by several developments, particularly the increased interest in international investing by sponsors. More and more funds are investing internationally, and those funds already invested overseas are increasing their international exposure. The most common reasons for the increased interest are risk reduction due to the diversification opportunities in international investing and the opportunity of achieving exceptional return.

Whatever their reasons for investing internationally, a key development is that U.S. sponsors are applying methods and approaches used by managers of their U.S. assets to their international assets and managers.

New Approaches

What are these approaches, and how do they influence the way international managers view benchmarks? Sophisticated U.S. sponsors typically have disciplined approaches to constructing their overall funds and typically institute some degree of risk control on the overall fund and on each manager. Sponsors often use multiple-manager analysis to look at the aggregation of the individual managers to determine the position of the overall fund. Another approach is for sponsors to develop customized benchmarks, or what we call normal portfolios, for managers whose style is not accurately reflected by one of the widely-followed benchmarks.

Sponsors using these methods do so to find a particular risk-and-return profile for each manager. Thus, a sponsor may hire several equity managers each with a slightly different expertise, for example, a growth manager, a value manager, a small capitalization manager, etc. Each manager may also have a slightly different benchmark, or normal portfolio, built especially for his or her particular area of expertise. The sponsor will then use the sum of the individual manager benchmarks to examine the fund's total exposure to the equity market. Funds are then allocated between the managers based on the desired exposure to each normal portfolio, which typically is the result of evaluating the risk-and-return profile of the overall fund.

The sponsor also monitors the risk of each manager's portfolio relative to the benchmark. If this risk, which is called active risk, is unacceptable to the sponsor, limits may be placed on the investment behavior of the manager. In general, U.S. sponsors have taken a more active role in the construction and management of their domestic portfolios, and this attention is spreading to their interactions with international managers.

Sponsors are taking a closer look at their international managers, how their portfolios compare to the benchmarks, and, thus, the active risk taken by international managers relative to the benchmarks chosen. It has yet to happen, but, in our opinion, some U.S. sponsors will begin to put limits on the investment aggressiveness of international managers by limiting the active risk the sponsor will tolerate from an individual manager. In view of this, sponsors and managers are both evaluating international styles and considering the use of normal portfolios to reflect their international styles in the same way that normal portfolios have been used for domestic portfolios.

Sponsors are also beginning to evaluate the risk-and-return profile of their *entire* international portfolio by combining individual manager portfolios into a multiple-manager analysis. While sponsors are concerned with the risk-and-return profile of individual managers, this is only one piece of the puzzle. The risk-and-return profile of the entire portfolio can only be gained through performing multiple-manager analysis to help sponsors know if the sum of the parts is a complete picture. All of these activities are, for the most part, accepted and widely used for U.S. domestic managers and, we believe, will increasingly be applied to international managers.

Quantitative Methods

The more structured approach discussed above typically involves using quantitative investment tools and analytics that, until recently, were not used or were not available for international investing. But quantitative methods are now becoming widespread and acceptable for international asset management and analysis.

As international asset allocation increases, we expect both sponsors and managers to increase their use of quantitative methods and models in the investment-management process and in the evaluation of international portfolios. Some of these processes include PC-based quantitative models addressing global and non-U.S. markets that have been developed and are currently being used for international investing.

Issues Associated with International Benchmarks

Most of the issues that have developed from the increased use of international normal portfolios are the result of perceived shortfalls in the traditional capitalization-weighted international benchmarks such as MSCI and EAFE. Often issues result from the increased complexity of constructing an international benchmark. It may be worthwhile to spend some time examining these issues.

International Benchmark Construction Complexity

An international normal portfolio is a portfolio like any other. It includes the names and holdings of a group of assets. However, one must decide on the assets to be included in the portfolio and the holdings of those assets.

The process of constructing an international normal portfolio is much more complex than the creation of a U.S. equity normal portfolio. An international normal portfolio must incorporate many markets and currencies. Instead of one decision that assigns the cash and market positions of the benchmark, numerous decisions must be made. Not only does one have to decide how much cash is to be in the portfolio but also in which currencies the cash is to be invested. The same goes for the equity part of the portfolio. Sponsors must decide which markets the benchmark will be invested in and how much will be invested in each market. Only then can the process of choosing the names of the assets to be included in the portfolio begin.

An additional problem is obtaining reliable, consistent data on the assets to be included, or to be considered for inclusion, in the benchmark. The lack of data may result in biases introduced into the benchmark that you cannot avoid, and is something that must be considered in the portfolio-construction process. For example, if data is unavailable for a market in which your manager routinely invests, the resulting benchmark will exclude the market and will not completely reflect the manager's style.

Either top-down or bottom-up approaches can be used to build normal portfolios. In a bottom-up approach, one would choose the assets in the normal and their weights and let the resulting country weights fall where they may. This approach, however, could result in some unusual or undesirable country weights. The more likely process would be a top-down approach in which one would develop country and currency weights and then develop the portfolios within each country. The top-down approach is used most often, but most normal portfolios modify only the country weights from one of the externally maintained benchmarks like MSCI or EAFE. The within-country portfolio remains the same as it is in the externally maintained benchmark, and still unaddressed are any biases or omissions associated with the within-country portfolios.

Benchmark Country Weighting

Probably the most important issue in the construction of an international normal portfolio is determining the weighting of each market in the benchmark. What is the "normal" country weighting? Financial theory suggests that a mean-variance efficient country weighting scheme that is based on risk and return predictions for each country and predicted relationships between countries should be used. However, the actual portfolio could be very different from the predicted efficient portfolio because of errors in the predictions of both risk and return. Furthermore, this methodology is rarely, if ever, used. What are some practical alternatives?

Two of the most popular country-weighting methods are capitalization and GDP weighting. Although these methods do not necessarily result in an efficient portfolio, capitalization and GDP weightings are easily calculated and are the methods typically used in the commercially available benchmarks that are widely recognized and quoted. Capitalization weighting weights each asset in the index according to its market capitalization. GDP weighting weights each country in the index proportionally to its gross domestic product, but the within-country portfolio is weighted according to the market capitalizations of each asset in that country.

The end result of the GDP weighting scheme is an index that significantly underweights Japan relative to a capitalization-weighted index. (This is the primary reason for creating the index.) It provides an economic reason for a benchmark that reflects the significant underweighting of Japan by most EAFE managers over the past few years. These are not the only alternatives for determining country weights, and many other alternatives exist.

"Vanilla" versus Customized Benchmarks

One question is whether a "vanilla," or widely used, externally maintained benchmark is adequate for sponsors and managers. Is the effort required to build and maintain a customized normal portfolio worth the benefits that are derived from its construction and use?

Suppose a manager or sponsor decides that a customized normal portfolio is worthwhile. The goal of a normal portfolio is to create a passive investment alternative to the manager's "style." However, if the manager is an aggressive market timer, as are many international managers, it is extremely difficult to identify a "normal" country weighting. In this instance, one probably would develop a weighting scheme based on subjective criteria such as conversations between the sponsor and the manager. Such discussions might include issues like South Africa and the manager's qualitative assessment of his or her normal position.

To Hedge or Not to Hedge

Another recent development is the interest in and use of currency-hedged benchmarks. Investment professionals are aware of the risk reduction that can be achieved in an international portfolio by hedging the currency exposures of the portfolio back to the sponsor's base currency. More and more sponsors are considering currency-hedged benchmarks. Unfortunately, the perceived risk reduction by hedging international benchmarks is not realized in practice.

The main emphasis of the sponsor's work should be reducing the risk of the overall fund. However, when evaluating currency hedging, sponsors usually look at the risk reduction of only the international portion of the fund. Typically, the risk reduction from currency hedging to the overall fund is quite small. Figure 1 shows the total risk, in variance, of a typical U.S. pension fund with increasing amounts of the fund invested in MSCI EAFE and currency hedged MSCI EAFE. With only 5 percent of the portfolio in international equities, the risk reduction to the fund by hedging the international investments is barely noticeable. It is our view that with only small portions of a fund invested internationally, the cost of hedging is not justified by the risk reduction achieved.

A Quantitative Comparison of International Benchmarks

So far, we've discussed the international indices currently in use and their recent modifications, custom country weighting schemes and currency hedging. What is the relative performance of the indices? Table 1 provides a risk comparison of several indices relative to the MSCI EAFE index. The figures in the table are predictions developed using BARRA's multiple factor risk model, the Global Equity Model. In the table we've provided the beta, residual risk and tracking error of each index relative the the MSCI EAFE index.

First, let's define the three risk characteristics provided in Table 1. *Beta* is the systematic risk coefficient of the portfolio relative to the benchmark. Beta measures how the portfolio will tend to perform based on the performance of the benchmark. If a portfolio has a beta of less than 1, its returns will tend to be lower in absolute magnitude than the returns to the benchmark. The converse also applies. If a portfolio has a beta of greater then 1, its returns will tend to be greater in absolute magnitude than the returns to the benchmark. Take, for example, a portfolio with a beta of 8.0. If the benchmark goes up by 10 percent, the portfolio will tend to go up by 8 percent. If the benchmark goes down by 10 percent, the portfolio will tend to go down by 8 percent.

However, the portfolio's performance will exhibit some differential performance around this beta trend line. This differential return is defined as *residual return*. The standard deviation of the residual returns is *residual risk*. The difference in the return to the portfolio and the return to the benchmark is defined as *active return*. The active return consists of residual returns plus *active systematic return*, which is the return which results because the portfolio has a beta other than 1. The standard deviation of the active returns is *active risk* or *tracking error*.

What do the figures in Table 1 tell us about the different international benchmarks? The FT EURPAC has a predicted tracking error relative to MSCI EAFE of 0.79%. This means that in two out of three years, the returns to the two will be within 79 basis points of each other (either positive or negative).

Figure 1 Adding Non-U.S.A. Equities to U.S.A. Fund

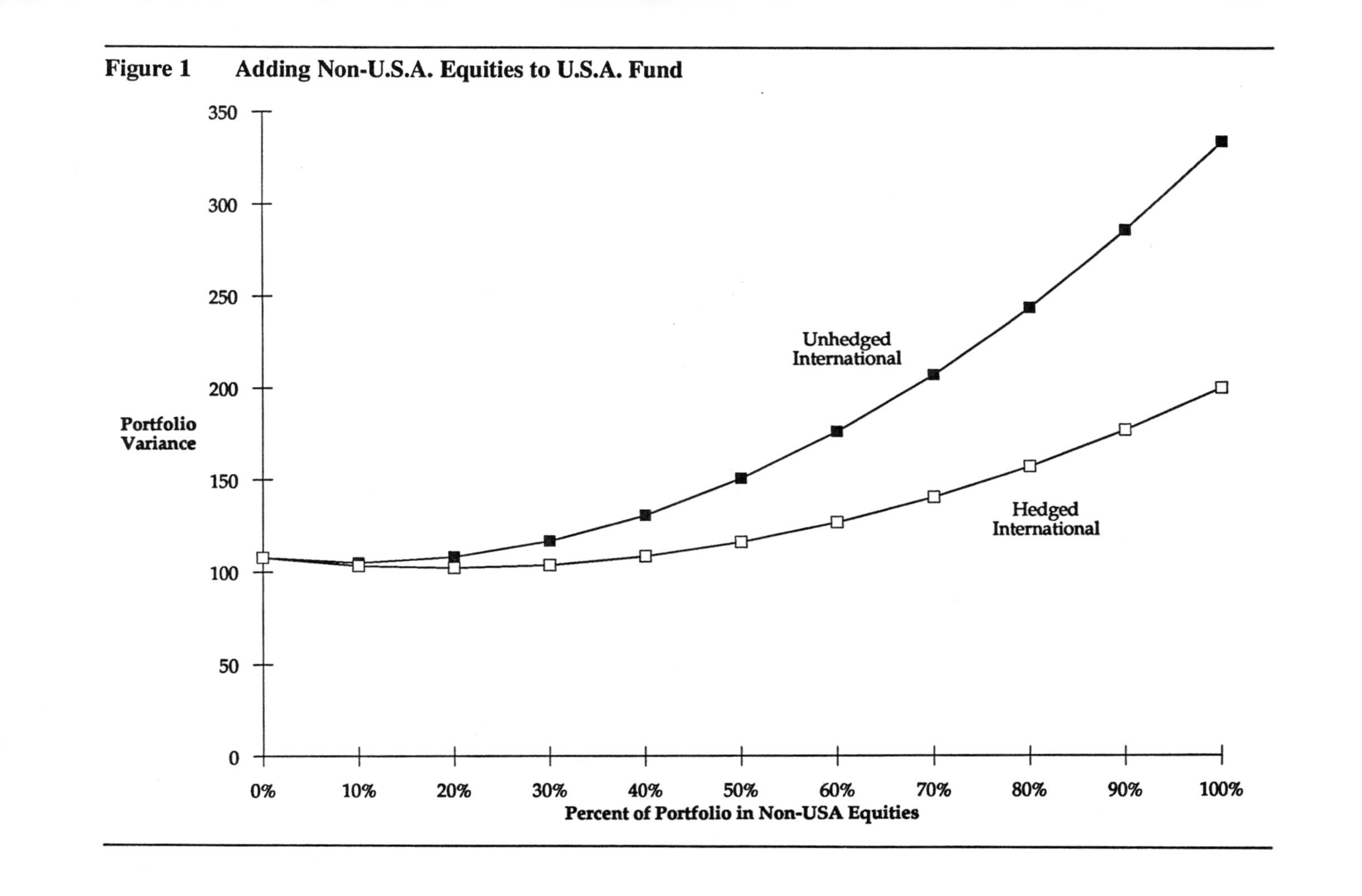

Table 1 A Characterization of International Benchmarks: Risk Comparison

FT EURPAC vs. MSCI EAFE	
β	1.01
Residual risk	0.78
Tracking error	0.79
MSCI WORLD vs. MSCI EAFE	
β	0.80
Residual risk	5.01
Tracking error	6.19
GDP Weighted MSCI EAFE vs. MSCI EAFE	
β	0.87
Residual risk	5.80
Tracking error	6.25
Currency Hedged MSCI EAFE vs. MSCI EAFE	
β	0.63
Residual risk	9.10
Tracking error	11.22

This relationship holds for the other three comparisons in Table 1; MSCI World vs. MSCI EAFE, GDP Weighted EAFE vs. MSCI EAFE and Currency Hedged MSCI EAFE vs. MSCI EAFE.

The predicted tracking errors between some of these different benchmarks are quite large, e.g., a 6.25 percent tracking error between GDP Weighted EAFE and Cap Weighted EAFE, and an 11.22 percent tracking error which results from currency hedging EAFE.

The issue of which benchmark is appropriate is a crucial one with major consequences, as these figures show. Yet, in many instances, the decision of which benchmark to use for asset allocation and performance evaluation gives short shrift to the differences discussed in this chapter.

Summary

We looked at international benchmarks currently in use and described some of their characteristics. We examined the traditional view of international benchmarks and some possible reasons for such a viewpoint, and some recent developments that have begun to alter the traditional view of international benchmarks. Finally, we looked at some current issues that we believe will affect the future use, construction, and characteristics of international benchmarks.

CHAPTER 12

The Development of Structured Portfolio Management: A Contextual View

William W. Jahnke
Vestek

Numbers serve to discipline rhetoric. Without them it is too easy to follow flights of fancy, to ignore the world as it is, and to remold it nearer to the heart's desire.

Ralph Waldo Emerson

None of us is as smart as all of us.

Wall hanging in James Vertin's office, Wells Fargo Bank, circa 1969, Unknown Author

The focus of conventional portfolio management is the identification and purchase of underpriced securities. Diversification by industry, and more generally, the management of risk exposures operates without well-defined quantitative control. The focus of structured portfolio management, by contrast, is the explicit management of portfolio risk and controlled exposure to factors that are associated with stock price movements. Structured portfolio management, although generally associated with indexing, encompasses common-factor tilting and the systematic implementation of stock valuation models. In structured portfolio management, the selection of individual securities is considered incidental to the process. Conventional portfolio managers select stocks; structured portfolio managers select portfolio exposures.

In this paper, the development of structured portfolio management is reviewed. The purpose is to assist the reader in understanding the business of managing portfolios quantitatively by describing some of the evolutionary path in the development of structured portfolio management. Theory, people, events, and technical issues are intermingled. There is no attempt to be complete. Many significant contributions have been omitted. There are many references to Wells Fargo in the paper. This is due to the important role that Wells Fargo personnel played in the development of structured management and the first-hand knowledge of the author of their roles.

The Origin

The origin of structured portfolio management lies in the development of a theory of portfolio risk and the subsequent development of a general equilibrium theory that describes the pricing of risky assets. Without the rapid development of computer data bases and processing capabilities in the late 1960s, the means to deal with the computationally demanding requirements of testing investment theories and quantitatively determining portfolio revisions would not have taken root in the 1970s and would not have challenged conventional investment practice in the 1980s.

The seminal work in the field of investment risk management was done by Harry Markowitz in 1952.[1] Markowitz was the first person to mathematically describe how portfolio diversification actually works. Although portfolio managers had understood that risk was reduced by increasing the number of investments in a portfolio and spreading investments across sources of risk, the process could not be mathematically defined. Since portfolio managers were not accustomed to quantifying risk, let alone return, there was initially little interest paid by the investment community to the great discovery.

Having dispensed with the problem of portfolio diversification, Markowitz offered a mathematical approach for determining the portfolio that offers the highest expected return for the risk borne. What is needed to make the calculation is the mean and standard deviation of expected returns for each investment being considered and the correlation of returns for each pair of investments. The set of portfolios offering the highest return for the level of risk is said to be on the "efficient frontier." Given their risk preference, all investors should choose a portfolio on the efficient frontier. To do otherwise is to accept lower return for the risk taken. If the investor's risk preferences could be described by a special set of parallel linear risk indifference lines, the "whole show" of determining the best portfolio for a given investor is solved by "quadratic programming," a nifty operations research tool.

[1] Harry Markowitz, "Portfolio Selection," *Journal of Finance* (March 1952), pp. 77–91.

Obviously, Markowitz was not your average professional investment manager. In fact he was not even in the investment business. Markowitz was a University of Chicago graduate student looking for a Ph.D thesis. While sitting waiting to see his advisor, he struck up a conversation with his advisor's stock broker, who was likewise waiting his turn. "And what do you do for a living?" Markowitz inquired. Markowitz found his thesis, and the investment management community got "modern portfolio theory."

There were several problems with modern portfolio theory. Investors were not experienced with making explicit, expected return forecasts, let alone quantifying their uncertainty. And the idea of forecasting thousands of correlation coefficients was enough to give a portfolio manager a migraine. As for quantifying a client's risk preferences as a set of risk indifference lines, nobody had ever thought about that before. Given the absence of large-scale computer processing capabilities in the 1950s, modern portfolio theory posed neither threat nor opportunity for investment managers, because the theory could not be practically implemented.

Almost a decade later Bill Sharpe, a Ph.D. candidate in search of dissertation, was working at the Rand Corporation, where Markowitz had taken up residence. Markowitz agreed to be an adviser to Sharpe on a thesis that would define a general equilibrium model for the pricing of risky assets expanding on Markowitz's celebrated 1959 book on portfolio selection.[2] The model was to become known as the Capital Asset Pricing Model (CAPM).[3] Sharpe was not alone in searching for a general equilibrium framework for the pricing of risky assets. John Lintner, Jan Mossin, Jack Treynor, and others made contributions in the mid-1960s.

According to Sharpe, who has emerged as the most visible spokesman, the uncertainty of returns for a stock can be broken down into two pieces. One piece of the uncertainty can be diversified away. The other piece cannot. Sharpe claimed that investors should not expect to be compensated for taking on risk that can be diversified away by merely combining stocks in a portfolio. The expected return for a stock should fall on a security market line that defines a general relationship between expected return and nondiversifiable risk.

Sharpe referred to nondiversifiable risk as "systematic risk" and the diversifiable risk "specific risk." A stock's systematic risk is determined by how the value of the stock is affected by system-wide events such as real GNP growth, inflation, real interest rates, and foreign exchange rates. Systematic risk captures only the risk of the stock that is correlated with the market. A component of the systematic risk of a stock or a portfolio was first referred to by Jack

[2] Harry Markowitz, "Portfolio Selection: Efficient Diversification of Investments," New York: John Wiley & Sons, Inc., 1959.

[3] William F. Sharpe, "Capital Asset Prices: A Theory of Market Equilibrium under Conditions of Risk," *Journal of Finance* 19 (September 1964), pp. 425–42.

Treynor as "beta." Technically, systematic risk is measured as beta times the risk of the market. A stock's specific risk is determined by uncertainties regarding the company's competitiveness independent of system-wide considerations. Corporations govern their exposure to both systematic and specific risk by their choices of product mix, capital structure, and by how well they select and manage personnel. It is to the quantitative measurement and control of these two forms of risk that structured portfolio managers devote themselves.

In a market where stocks are not priced to fall on the security market line, there is an opportunity to create portfolios offering superior return for the risk borne. Competitive market forces should naturally develop to eliminate such opportunities. By increasing the price for a stock whose expected return is above the line, the expected return is bid down to the security market line. Conversely, overpriced stocks whose expected returns plot below the security market line should fall in price, bidding up their expected returns until they are positioned on the security market line. By continuously monitoring company prospects in looking for investment opportunity, investors should adjust stock prices to maintain the stock's position on the security market line. A market in which all stocks are fairly priced is called an *efficient market*. The theory that all stocks in a market should be fairly priced is called the "Efficient Market Theory."

In 1964 it was evident investors were not thinking explicitly about CAPM in setting prices and building portfolios, but, perhaps in a less formal way, investors were behaving in a manner consistent with the theory. Most professional investors thought about earnings growth prospects in assessing the appropriateness of the price-to-earnings ratio, and they considered portfolio concentrations when selecting stocks for portfolios. In these deliberations consideration of returns and diversifiable risk were present.

Not surprisingly, the academic and investment management communities did not embrace one another on CAPM implications for a fairly priced stock market. The issue of market efficiency had more than academic truth and beauty riding on it. At risk was the financial livelihood of a good part of the investment community. If stocks were fairly priced, why should investment managers be paid exorbitant management fees? Why should investment managers trade stocks to improve investment performance? Why should brokers earn large commissions for providing an unnecessary and costly service? The answers were all too clear.

Fortunately for the investment community before the 1960s, nobody had much evidence one way or the other on just how efficient the stock market was. With the introduction of the computer, the academic community had the tool to making serious studies on what they called "random walks," "martingales," "fair games," and "efficient markets." The academic work was divided into what

Gene Fama later called weak-form, semistrong-form, and strong-form tests of the efficiency of the stock market in setting prices.[4]

According to Fama the voluminous weak form tests that evaluated, and in some cases attempted to devise, trading strategies based solely on historical price and return sequences movements were unable to find sufficient departure from the efficient market hypothesis "to declare the market inefficient." The semistrong form tests, which studied the effects of public announcements such as stocks splits, dividends, and earnings, likewise produced little in the way of investment opportunities, after adjusting for transaction costs. The study by Michael Jensen of 115 mutual funds over the 10-year period 1955 through 1964, which arguably is a strong-form test if one believes the managements of mutual funds are operating with nonpublic information, was perhaps the most damaging of all the early work.[5] Using the CAPM model to risk adjust returns, Jensen found that 58 of 115 funds produced returns below the security market line, even when load charges and all expenses excepting brokerage cost were ignored. The average deviation of 10-year returns from the security market line was minus 2.5 percent. All costs in the average fund underperformed by 14.6 percent with 89 of 115 funds underperforming. If there were funds that consistently produced above-average results, Jensen evaluated the subperiod performance. He found that above-average returns in one subperiod were not associated with above-average returns in subsequent periods. Although the results applied to only one segment of the investment management business, Jensen argued that they were striking evidence of the stock market being efficient. Although these early stock-market studies supported the efficient market theory, they carried little weight at the time with the purchasers of investment management services. The pension-fund market could hardly be expected to accept studies of the mutual fund industry incorporating a new theory of price formation to establish its point as relevant to the management of pension fund assets.

The idea of managing portfolios to match the market returns would have remained unfulfilled had it not been for the use of the computer in the measurement of pension-fund performance. It was Merrill Lynch and A. G. Becker who got into the business of measuring portfolio performance based on the results of a study commissioned by the Bank Administration Institute. What better way to establish the existence of superior managers and to get the laggards to try harder, and incidentally produce more turnover and higher profits for Wall Street, than to measure performance. Not surprisingly, the results were similar to those found by Jensen. The average pension equity manager produced returns below the market averages. After all, the managers controlled an increasing share of the assets

[4] Eugene F. Fama, "Efficient Capital Markets: A Review of the Theory and Empirical Work," *Journal of Finance* 25 (May 1970), pp. 383–423.

[5] Michael Jensen, "The Performance of Mutual Funds in the Period 1945–64," *Journal of Finance* 23 (May 1968), pp. 389–416.

being managed and they had to pay Wall Street for brokerage services. The academics, never at a loss for an eloquent explanation, pointed out to institutional investors that the stock market is a zero-sum game, less the cost of playing. Not everybody can be a relative winner. For every winner, there is generally a loser in relative performance, before the cost of playing. (The costs being brokerage and management fees).

Although some conventional managers were willing to concede the zero-sum game argument, they had a great problem accepting a lack of consistency in performance, which the data likewise exposed. According to the evidence, good investment performance took a random walk. Among pension-fund managers, there were no more managers with a successful string of investment successes than would be produced by chance.

Index Funds

Far away from Wall Street, on the West Coast, the Wells Fargo Bank's Management Sciences Department, under the direction of John McQuown, had been seeking a new road in investment management. The goal of the Management Sciences Department was to assist the bank's trust investment division, known then as the Financial Analysis Department (FAD), in exploiting the institutional market by introducing rigorous quantitative investment management. Unfortunately, the security analysts and portfolio managers in FAD didn't think too much of higher mathematics or the unproven theories being thrust upon them; although the head of the department, James Vertin, had developed, in part through his tumultuous association with McQuown, a high level understanding and appreciation for the "new technology," as he called it, and its implications for investment management. In an internal memorandum to staff dated November 19, 1970, Vertin writes, "While only time and experience will tell, it may well be that we are very much in on the ground floor of a development which can mean big things not only for Wells Fargo Bank, but for FAD and each staff member individually."

Because of its close ties with academia, which included sponsoring academic forums in the bank's board room, the Management Sciences Department was contacted regarding a possible piece of investment management business. One well-connected member of the Samsonite Corporation had completed his studies at the University of Chicago and was surprised to find the Samsonite pension fund was invested in mutual funds. Feelers were put out to see if anyone in the business was managing pension-fund assets in a "theoretically proper" manner. Jim Lorie, Merton Miller, and Gene Fama, professors at the University of Chicago, knew of the work being done at Wells Fargo and suggested Samsonite call Wells. Discussions with the bank resulted in the Management Sciences Department defining an investment strategy and the Financial Analysis Department managing the strategy. The strategy Management Sciences came up with was to create a portfolio whose objective was to track the performance of an equal-dollar-weighted New York Stock Exchange Index. The choice of the NYSE was largely based on the fact that it was a broad-based index, and by equal-weighting it, the strategy held the prospect of high returns due to its relatively high level of systematic risk. The portfolio to be managed was, arguably, the first structured portfolio.

In July 1971, with a $6 million contribution from the Samsonite Corporation, the first sampled index fund was activated. Because the established portfolio management group in the Financial Analysis Department did not want to have anything to do with managing a "passive" portfolio, where a computer decided what stocks to buy, it was left to Bill Fouse, who had recently joined FAD, and his nascent Investment Systems Group to manage the strategy using software developed by the Management Sciences Department's Wayne Wagner and David Bessenfelder.

The experience gained with the Samsonite account encouraged the Management Sciences Department to enter the investment management business by offering services consistent with CAPM. Besides, they had worked long enough with too little success in reforming the bank's Financial Analysis Department. In collaboration with two academics, Myron Scholes and Fischer Black, they formulated a new investment product using modern portfolio theory and empirical research using data supplied by the Center for the Research in Security Prices (CRISP). A leveraged low-beta mutual fund, to be known as the Stagecoach Fund, was to be the first investment product. The idea was to get pension-plan sponsors to invest in a collective fund that would borrow short and invest in a passively managed portfolio that would track the performance of low-beta stocks. Scholes and Black's research had revealed that high-beta stocks had not produced investment returns commensurate with their systematic risk. This stock-market pricing inefficiency they called the "alpha effect." By leveraging a low-beta portfolio, superior returns could be earned for the level of risk taken.

The idea of using leverage was already established as the theoretically proper approach to managing portfolio risk. Yale Professor James Tobin had demonstrated that the choice of stocks held in a portfolio was a separate decision from the decision of how much risk to take.[6] An investor would always own some combination of the stock portfolio offering the highest return per unit of risk and lend or borrow a riskless asset. This important investment concept was called the "Separation Theorem," for which Tobin received in part his Nobel Prize in economics. Sharpe demonstrated with CAPM that the market portfolio should be the portfolio that provides the highest return per unit of risk and that

[6] James Tobin, "Liquidity Preferences as Behavior Towards Risk," *Review of Economic Studies* (February 1958), pp. 65–86.

all investors should hold some combination of the market portfolio and the riskless asset.

The marketing team in the Management Sciences Department thought offering a leveraged index product a terrific idea. Not only did the theory recommend leverage as the proper means for achieving above-average risk exposure, but the research at CRSP had detected a possible chink in the efficient market theory, which made the idea of leveraging a low-beta portfolio even more desirable. Besides, it seemed to them that the idea of offering a plain index product was not likely to capture the imagination of pension-plan sponsors.

After two years of intensive product development and marketing, the project had identified a handful of prospects, including Greyhound and Illinois Bell, when the project was adjudicated out of business. Back on the East Coast, the Supreme Court had sustained the suit brought by the Investment Company Institute (ICI) against the Comptroller of the Currency. In *ICC* v. *Camp*, the charge was that the comptroller's permission to allow CitiCorp to distribute collective fund participation through its branch network was in violation of Glass-Steagall Banking Act of 1933, which separated commercial banking from investment banking. Wells Fargo Bank management, on the advice of outside legal counsel, decided to retire from the field. The Management Science Department would not be an investment manager.

James Vertin had watched with interest the efforts of the Management Sciences Department to enter the investment management business. Unwilling to see a piece of business pass by, Vertin's group stepped up to the plate and offered to set up a collective fund to track the performance of the S&P 500 Index. Up until that time FAD, later to be renamed Wells Fargo Investment Advisors (WFIA), had a large number of personal-trust and investment-counselling accounts totalling $2 billion, but only $75 million in major institutional business. Matching Wells Fargo Bank's contribution of $5 million from its own pension fund, Illinois Bell became in 1973 the first client in the Wells Fargo S&P 500 Index Fund, which was a common trust fund set up by the bank's Trust Department. The S&P 500 Index was chosen instead of the NYSE, because it had gained respectability as the appropriate institutional performance comparison. It was also easier to index than the NYSE. The S&P 500 was capitalization weighted, representing 65 percent of the capitalization-weighted U.S. equity market, and it could be replicated with only a $25 million investment.

Since the initial funding of $10 million was not enough to buy all 500 stocks in proportion to their weight in the Index, a sampling strategy was devised that would buy the big capitalization stocks in the index in proportion to their weight in the index, and would sample the small stocks. The small stocks were sorted into the cells of a matrix consisting of historical beta and economic sector classifications. The idea was to buy equal weighted positions of small stocks to match the capitalization weighted cells of the matrix. As money was added, taken away, or as stock prices moved, the program would redefine big stocks, match their portfolio weight to the Index weight, and match the portfolio weight in the matrix to the Index's weight of small stocks in the matrix. As the dollar amount in the portfolio approached $25 million, the number of portfolio holdings would approach all 500 stocks in proportion to their weight in the Index.

The first S&P 500 Index trade list was generated from a computer program developed by Virginia Chung. It was sent by messenger to Salomon Brothers in San Francisco, who traded it over a several-day period, acting as agent and charging the then normal commission rate of 28 cents per share, which was roughly a 1-percent commission. Over the next three days, confirms poured into the Bank. The Bank's trust accounting group failed to get the trades posted before the night stream cutoff. It was reported that the Security Clearance Department suffered apoplexy clearing the trading program. A new era had begun, which would, and still is, reshaping trading and back office operations.

One of the problems in not owning all the stocks in the Index in their proper weight is that the performance of the portfolio will deviate from the performance of the Index. The deviation of performance is a manifestation of what Sharpe calls specific risk, which indexers prefer to call *tracking error*. Tracking error is the periodic deviation in return between a portfolio and the target index. Technically, specific risk and tracking error are the same when the portfolio beta is 1.0. Tracking error is usually reported as a standard deviation of returns. The less quantitatively oriented prefer the unsigned average deviation of portfolio returns from the index returns. Full replication index funds track their indices within several basis points per month. Sampled index funds, by their nature "dirty" index funds, tend to produce significant tracking errors. The first index fund was expected to experience tracking error because it was forced to sample.

In general, most indexers now try to buy all the stocks in the target index. Full replication indexers don't like accepting specific risk for which there is no expected compensation. There is also concern that the method used to sample the index may result in buying stocks that are overpriced, offer lower expected return because they are in general less risky, or will reflect a bias on the part of the portfolio manager that exposes the portfolio to higher degrees of specific risk than are necessary.

One of the early indexers and its vendor providing the indexing software promoted the idea of sampling over full replication. The arguments they used in favor of sampling were lower transaction and custodial costs. By using optimization, they claimed they could create a portfolio with as few as 100 stocks that would perform along with the S&P 500 with an acceptable tracking error. The trade-off of holding fewer stocks to lower transaction costs is difficult to accept. Trading costs are generally a function of cents per share plus price impact. Buying fewer stocks does not reduce the number of shares purchased, while concentrating purchases in fewer stocks increases price impact. However, custodial charges are often based on the number of stocks held in a portfolio and the number of trade tickets processed. In deciding how many stocks to hold in an index fund and the minimum transaction size in an index fund trading program, the tradeoff between tracking error and custodial costs should be evaluated. Since custodians price differently, it is worth comparing custodial pricing structures.

A number of the clients who have had their index funds managed using sampling techniques have been disappointed in the tracking. This is due to bias introduced into the portfolio based on the method used to select the stocks. Letting the portfolio manager choose often introduces bias. Portfolio managers are naturally drawn to well-recognized stocks, which appear to be safe investments, and which have done well in the past. Unfortunately this is not a prescription for unbiased future performance. Some indexers running sampled portfolios relied on portfolio optimization programs to find the portfolio with the lowest estimated tracking error for the number of holdings in the portfolio. This is initially done using specific risk measurements based on historical returns, that is until Barr Rosenberg introduced his now famous Fundamental Risk Model.

Barr Rosenberg was consulting for Wells Fargo's Management Sciences Department in the late 1960s on bank operations research applications. He became interested in the investment applications and began consulting to WFIA. Barr was intrigued with the idea of using fundamental factors to improve on the estimates for portfolio systematic and specific risk.[7] Barr Rosenberg and Walt McKibben, financed by a National Science Foundation grant, employed a multifactor econometric system to estimate the covariance of returns to factors and in 1973 published their method for estimating portfolio systematic and specific risk. Observing an opportunity to bring institutional investment management a superior methodology for controlling and accounting for portfolio risk, Barr established his consulting firm, BARRA.

The reason for going through the exhausting data-base creation and the elaborate econometrics was to reduce the estimation error in the estimates for beta and specific risk. Barr regressed monthly returns against accounting, fundamental, and technical factors. Rosenberg argued that risk estimates using a fundamental risk model reduced bias in the covariance matrix from chance events. Even so, there remain serious problems with using past data to estimate risk. If, for some reason, the collection of events in the period measured are not indicative of future probable events, estimates of future portfolio risk are misrepresented. The problem is compounded when a portfolio optimizer is used to search out stocks that produce low portfolio estimates of total risk or low tracking error with a target index. The optimizer will purposely seek stocks with the lowest

[7] Barr Rosenberg and Walt McKibben, "The Prediction of Systematic Risk and Specific Risk in Common Stocks," *Journal of Finance and Quantitative Analysis* 8 (March 1973), pp. 317–33.

covariances and specific risk measurements, which are likely to be downward biased. The result is an underestimation of portfolio tracking error. Realized tracking errors can run 30 to 40 percent higher than estimated tracking errors for indexers using portfolio optimization to construct sampled index funds, if the risk estimates are not adjusted.

Why would an indexer advocate sampling strategies? After all, sampling makes the process of running index funds more complicated. Sampling requires more sophisticated computer systems. Sampling requires the portfolio managers to be more technically proficient. Sampling opens up the prospect of poor tracking. One explanation is that sampling is necessary when portfolios are not large enough and commingling with other portfolios is not available or otherwise acceptable. Most indexers endorse full replication of the index and participation in commingled index funds. Only in the cases of insufficient funds, problems of market liquidity, round lotting, or a desire to control the number of trades in a portfolio update, would an indexer accept not owning a pro rata share of the index.

In the case of the S&P 500 Index, Standard and Poor's provides a service that defines and updates the holdings and the number of shares outstanding for each stock in the index. Computer index matching programs produce and update portfolio holdings. Portfolios created by a matching program can be expected to closely track the return on the index, assuming there are no major changes in the index. The matching of index fund returns to its target index, assuming no revision to the index, can be thought of as *intrinsic* tracking. Intrinsic tracking error is in the control of the portfolio manager. Full replication index funds generally have intrinsic tracking errors of less than 1 basis point per month. Depending on the number of holdings and the level of diversification, sampled index funds generally have intrinsic tracking errors in excess of 20 basis points per month and in some cases much higher.

Other sources of tracking error that require ongoing action by the manager can be thought of as *exogenous*. The exogenous sources of tracking error include dividend reinvestment, additions and deletions from the index due to merger or index revision, and changes in shares outstanding due to corporate stock repurchases or issuance. In the absence of exogenous sources of tracking error, the manager of a fully replicated index fund could go on holiday, and the index fund would continue to track along with the index. Changes in individual stock weighting in the index due to price movements would be mirrored automatically in the index fund.

One problem in managing S&P 500 Index Funds has been obtaining an accurate measurement of the total return of the Index to serve as a benchmark for evaluating how well the Index Fund is being managed. Surprisingly, there has been a wide divergence in S&P 500 total return calculations. In some years the total return for the S&P 500 reported by various industry sources has diverged by as much as 50 basis points. Part of the problem in return calculations has been the reinvestment of dividends. The S&P 500 returns reported in the newspaper are price change only. Because the daily return on the Index is not materially affected by daily dividend income, the convention in the business has been to ignore dividends when calculating the daily return. Organizations managing index funds or measuring and comparing investment performance were required to make the total return calculations themselves. In some cases, dividends were assumed to be reinvested monthly or quarterly. The more accurate methods accrued on ex date and reinvested on ex date, pay date, or in the most accurate case, five business days prior to pay date.

The treatment of additions and deletions from the S&P 500 is another contributor to return discrepancies in calculating the return for the S&P 500. The most accurate methods add and delete stocks at the time S&P makes its index changes, which occur on Wednesdays at the market close. The least accurate S&P 500 return calculations assume rebalancing once a month or once a quarter. A similar problem exists for stock buybacks or issuance where the monthly or quarterly assumptions are used. In some cases, Standard and Poor's is not current on the number of shares outstanding, which poses the question of whether to use S&P published data or more current information from whatever source. Calculating S&P 500 total return daily requires updating share changes before Standard & Poor's reports them.

Once a reasonably accurate measurement of the S&P 500 is available, the problem confronting the index manager is tracking the return on the index. In the case of an S&P 500 Index Fund whose objective is full replication with minimum tracking error, there are several problems. Prior to October 1989, Standard & Poor's would announce changes to the index after the close of the market on Wednesdays. Because index funds could not rebalance until Thursday, they had to absorb the overnight risk of adverse price movements. Not surprisingly, on Thursday when indexers went to buy, the prices were usually higher by several percent, and when they sold, the prices were lower by several percent. Some indexers, in order to reduce the performance disadvantage, play the stocks over subsequent days trying to decide the best time to transact, looking for some post announcement price reversal. With pre-announcement of index revisions, the price effect of index revisions has an opportunity to work into the prices used by the index in calculating the return on the index.

The merger activity in some years has produced significant turnover in the index accounting for most of the sales turnover. In recent years the S&P 500 Index turnover has averaged 4 percent. Stock repurchases and S&P restructuring have added yet more turnover. The large purveyors of indexing feel compelled to maintain very close tracking to the defined index for marketing reasons, even though it has resulted in turnover costs that are not offset by expected return pickup or total risk reduction. Some pension plan sponsors who have decided to manage their index funds in-house have elected to reduce or avoid sales turnover. Since the tracking error in an index fund has a negligible impact on the total risk of the fund, while the brokerage costs for trading are real, they argue against sales turnover for rebalancing, favoring some level of tracking error.

Generally speaking, an index fund should be reviewed at the frequency with which index revisions are published by the keeper of the index. In the case of the S&P 500, the frequency is weekly. In the case of other indices, it is monthly, semiannually, or annually. Events such as tender offers are handled as they occur and are signalled by the custodian. Brokers are often very helpful in providing information and insight in determining the appropriate response to tender offers. In some cases informal networks of in-house index fund managers exchange information and opinions on tender offers and other issues confronting indexers.

The reinvestment of dividends is one of the largest concerns for index fund managers. Although the industry convention is to consider dividend reinvestment in calculating the return for an index fund, there is no economic theory requiring it. Dividend payouts from S&P stocks are not necessarily reinvested back in S&P stocks by investors receiving them. From an accounting standpoint, on ex date an index fund could choose to make a disbursement of the pay date right to dividends. Periodically, accumulated dividends could be reinvested in the index fund and handled as a contribution. The time-weighted rate of return of the S&P 500 Index Fund with periodic contributions of dividend could then be compared with the S&P total return calculation.

When it comes time to rebalance the index fund or to reinvest dividends, proceeds from tender offers, or contributions, a portfolio rebalancing program is run. Just because the objective of the index fund is to own all 500 stocks in the right proportion, it does not follow that the best thing to do is to buy 500 stocks in each purchase program. Stocks should be purchased and sold in round lots. The portfolio manager must evaluate the tradeoff of how many different stocks to buy in relation to tracking error and custodial costs of settling the trade program. The more trade tickets, the higher the custodial costs. Index fund trading programs control minimum transaction size to control the number of tickets. Tracking reports project the level of portfolio tracking before and after a proposed trading program. In many cases the number of trades in a program has minimal impact on portfolio tracking. So why produce a lot of tickets!

Index fund trading has undergone major changes from the time the first S&P Index Fund was traded. The first trade list was passed to one broker who acted as agent, trading the program over several days. Because the S&P Index Fund was not an instant marketing success, it was not until several years later, after the 1973–74 market break, that contributions began to flow into index funds, when institutional investors suddenly got religion about controlling portfolio risk. With the onset of negotiated commissions in 1975 came the opportunity to significantly lower index fund trading costs. This made trading the index a low cost proposition providing tremendous impetus to indexation.

In the past, brokers had to protect themselves from information that the investor might have regarding the value of the stock not reflected in market price. Brokers accomplished this by charging large commissions and setting large bid/ask spreads, which permitted them to transact out of their inventory or to short stocks without the necessity of knowing their value. Index trades on the other hand were seen as containing no company specific transaction risk. Therefore, brokers only need be concerned with market-wide price movements. WFIA's Tom Loeb understood the risks that brokers face having worked previously for a major broker. He was able to interest several major brokers in bidding as principal on a package of "informationless" trades. Taking a page out of the block trading book, these brokers were willing to put up their capital to guarantee the trade program at closing prices.

Index fund traders and their counterparts in brokerage firms initially developed a symbiotic relationship, which is atypical in most buy-sell side relationships. There were few brokers willing to risk their capital and only a few organizations managing index funds. There was an early recognition that the business was relationship oriented. There was little to gain and a lot to lose if one side or the other came out a loser in index trading programs. Both sides needed to be there the next day. Because of the desire of the indexers to track the index, indexers were interested in obtaining closing prices, because the calculation of the unit values of the index funds was based on closing prices. To transact at different prices during the day opened up the prospect of tracking error. The indexer did not want the broker to transact at the closing price, because that would exert pressure on stock prices. Because the indexers time-weighted return calculation used closing prices, the price impact was hidden from indexers' client.

What emerged was a framework in which the broker guaranteed the closing price as principal for a commission. The size of the commission depended on the rules established for trading. If the broker was allowed to trade the program up to one hour before the close of the market, a lower commission was charged than if the broker had to wait until the next day to trade. Some programs had the broker rebating a percentage of any trading profit that was made by the broker. Not surprisingly on some of the programs brokers lost their shirts if the market went against them. Some of the brokerage trading desks thought they could call the market. When they couldn't, red ink flowed. In the early stage of package trading, there was a high degree hope if not confidence that the broker would play it straight with package trading. Although the broker in many programs was required to trade prior to one hour before the close, the broker could trade for his own account at any time. With new entrants and catching-up for prior losses, brokers began cutting corners in index trading. It was difficult for indexers to police the brokers. This was especially true when the brokers could play the stock index futures to hedge their risk, or front run the programs. When indexers put programs out to bid, the losing brokers could position themselves.

In the mid 1980s, trade execution programs were developed to measure the cost of trade execution. Studies of index trading programs indicated that the closing prices were being marked up when indexers went to buy. As a result index trading practice in recent years has been moving away from using guaranteed programs.

The development of the stock-index futures did more than provide the broker a method for hedging index trade program risks. Now the indexer had a new tool to play the market portfolio. Combinations of Treasury bills and S&P 500 Stock Index Futures could be used to replicate the performance of the S&P 500 Stock Index. A product called *index arbitrage* was created, which involved selling the future when it was overpriced and buying the stocks in the index to produce a very low risk investment that offered higher returns than Treasury bills. The aggressive implementation of index arbitrage and the buying and selling of packages of stocks was quickly blamed for producing instability in the stock market. The foes of indexation seized the opportunity to attack indexing through its association with package trading. What was a natural and desirable process of linking the cash and futures markets through arbitrage, which provides the means to effective hedge market risk exposure, was characterized as destroying confidence in the stock market due the "mindless" trading practices of indexers and index arbitraguers.

The controversy hit full stride on October 16, 1987, when the Dow Jones declined 20 percent. An innovation in structured investment management known as *portfolio insurance* was linked to the crash. Portfolio insurance, the brain child of two U.C. Berkeley professors, Mark Rubinstein and Hayne Leland, had been successfully marketed to a number of plan sponsors. The idea was to protect against large portfolio losses by selling stocks or the index futures in declining markets. The amount of selling was determined by complex mathematical formulas devised by the professors. As the market declined, the portfolio insurer would continue to sell stocks until the portfolio insurance program stopped selling. The idea and market implications of selling low and buying high were troublesome to some of the quantitatively oriented in the profession.

When the market headed south in mid-October, the portfolio insurance programs began to bail out of stocks and to sell the futures when they were not too far away from being fairly priced relative to the underlying stocks. The Brady Commission was formed to investigate the crash and to make recommendations to bring stability and rationality back to the market. The Brady Commission stopped short of what some of the industry had hoped, which was to recommend policies that would eliminate portfolio insurance, cash arbitrage, package trading, and (why not) indexation. Ironically, because of a failure of the futures market to track its fair value by selling at a huge discount, several of the institutional managers could not hedge their positions, and performance floors were penetrated. As a result of the failure of some practitioners to produce the promised protection in declining markets and the heat associated with participating in a practice that, by its nature, is associated with adding fuel to a declining market, the number of plan sponsors employing portfolio insurance has fallen dramatically.

Some large sponsors in recent years have taken the management of their index funds in-house. This trend is likely to continue. Using data bases and computer programs provided by outside vendors, sponsors have been successfully producing index-like results with minimal internal resource requirements. The growth in the number of indexers from a few major banks to dozens of institutional and in-house managers in recent years, has created an opportunity for indexers to trade positions among themselves, creating market liquidity away from Wall Street. Some of the very large indexers routinely cross index positions between their clients. Several crossing networks are now operating, which bypass the New York Stock Exchange. Meanwhile Wall Street has improved index trading opportunity with Superdot, which provides the opportunity to trade sizable baskets of stocks through the New York Stock Exchange. The development of additional basket products, indices, and index futures will undoubtedly occur, permitting the broadening and increased cost-effective management of index funds.

The S&P 500 was the logical place for structured portfolio management to take root. The S&P 500 provided a manageable benchmark of liquid stocks whose returns could be replicated with a small tracking error. Not represented in the S&P 500 was 35 percent of the equity-market capitalization consisting of approximately 4,500 medium to small capitalization companies. Because pension funds were also invested in the smaller stocks not represented by the S&P 500, good performance of small stocks relative to big stocks in a given period would influence how the median portfolio manager performed relative to the S&P 500. Periods in which small stocks performed better than large stocks were better for active managers but bad for S&P indexers. Indexers concerned with implications of falling prey to a period of small stock performance had several choices. The indexer could manage an index fund representing a larger percentage of the market or set up a special index fund to represent the 35 percent of the market not represented in the S&P 500. Because the S&P was so well entrenched as the institutional benchmark, the latter course was initially chosen. The non-S&P 500 Index was created to match the performance of the stocks not included in the S&P 500. Because of the large number and illiquidity of the stocks, a sampling strategy was employed similar to the one used with the first S&P 500.

The development of the non-S&P 500 Index Fund placed active management performance at a greater disadvantage than before. The combination of the S&P and the non-S&P Index funds provided tougher competition year to year, because it leveled out the effects of big or small stock years in the market. The existence of the S&P 500 Index Fund as a big stock year strategy and the existence of the non-S&P Index Fund as a small stock year strategy were alternately tough competition for active managers.

The S&P 500 Index Fund can be thought of as factor-tilted toward large capitalization stocks. The non-S&P 500 Index Fund is factor-tilted toward small capitalization stocks. Investing the right percentage of assets in each fund produces a combined equity exposure representative of the total U.S. equity market. In initially choosing the S&P 500 to represent the U.S. equity market, the bias or tilt toward large capitalization was incurred. Since smaller capitalization stocks should be expected to produce higher returns over time to compensate for higher risk and lower liquidity, the S&P 500 Index Fund should be expected to provide less return than the non-S&P 500 and lower return than the median actively managed portfolio before transaction costs and management fees. The acceptance of a lower gross return was considered acceptable at the time in light of the beneficial features the S&P 500 Index offered.

The growth of index funds has been phenomenal. Index funds grew from $6 million in 1971 to $10 billion by 1980 and to $250 billion by 1990. Today index funds represent 30 percent of the institutional equity assets and, according to some visionaries, are likely to exceed 50 percent by the end of this century. This prospect is not without its detractors!

In recent years the S&P 500 has been a very hard benchmark for active managers to beat. Less than 30 percent of active managers outperformed the S&P 500 during the past five years before management fee's. These results are even more startling when the fact that managed accounts that have been terminated are not included! At fee schedules that one prominent consultant called an "embarrassment," the movement toward indexation has significantly reduced the potential earnings for investment managers and Wall Street. Among charges of being "un-American" and "mindless investing," a growing number of corporate and pension plan sponsors are settling for average and placing at least some of their assets in index funds. According to one sponsor wishing anonymity, "it is better to have a high probability of being average investing in index funds than to hire a stable of active managers and have a high probability of being below average." Questions of the efficient allocation of capital and the ultimate destruction of the capitalistic system as we know it have been raised. These concerns are particularly amusing given the misallocation of capital associated with the leverage buyout craze of the 1980s and the fact that very few of the critics can speak authoritatively on the subject of the cost of equity capital or use it in any form in their stock-selection processes.

Common Factor Investing

The efficient-market advocates would naturally choose an index fund that was broadly representative of the U.S. equity market. But what about those who believe or would like to believe the market contains pockets of inefficiently priced stocks? There are academics and investment managers who believe inefficiently price stocks can be identified by simple attributes such as p/e, yield, and market capitalization. Attributes that can be used to describe stocks have been accorded the label *common factors*. Other common factors such as forecasted earnings growth rates, changes in consensus earnings forecasts, analyst coverage, and relative strength all have advocates who believe these common factors, individually or in combination, routinely identify mispriced stocks. Even the early studies identified historical beta as being mispriced. Numerous studies on the relationship between investment performance and p/e ratios, conducted as far back as the 1950s, found abnormal return potential in low p/e investing, which was conveniently ignored or dismissed by efficient-market advocates. Interestingly, in many cases among investment managers with good track records, most if not all of the superior performance can be explained by exposure to one or more common factors. The question naturally arose whether it was possible to manage portfolios that were run like index funds but maintained desirable common-factor exposures.

It was in the mid 1970s when the opportunity to manage factor-tilt portfolios arrived. The successful tracking of S&P 500 returns by the S&P 500 Index Fund had demonstrated that computer-based investment strategies could be successfully implemented. In 1974 Congress passed the Employee Retirement Income Security Act (ERISA), which provided a legal defense for both index and common-factor-tilted portfolio management. Up until ERISA, investment managers had been governed by the *prudent man rule*, requiring each investment in a portfolio to stand on its own merits by current industry standards. Buying a stock or a basket of stocks based solely on the fact that they had low p/e's without support from traditional fundamental factors, was guaranteed to run afoul of the prudent man rule if put to a legal challenge. Since most factor-tilt strategies ideally buy as many stocks as possible having the desired factor exposure, it was not practical to do the basic research and develop the conventional defense for stock purchases. ERISA introduced the *portfolio standard* which held that the basis for evaluating the appropriateness of any investment was its contribution to the portfolio as a whole. With ERISA a defense could be made for owning a diversified portfolio of low p/e stocks, where many of the individual holdings would be considered imprudent investments by the earlier standard. Even with a new legal basis for common-factor-tilt investing, investment managers and pension-plan sponsors were not quick to jump at the opportunity.

One of the common factors that had been identified as offering superior returns over time was high dividend yield. A Wells Fargo Index Fund client, who had a desire for high current income and was aware of the research on the superior historical performance of high-yielding stocks, inquired about the possibility of tilting their index participation toward high yield. The idea of offering a yield-tilted portfolio strategy was especially attractive to Wells Fargo, because its work with the Dividend Discount Models indicated that above-average-yield stocks offered superior expected returns. A portfolio strategy was designed to offer 1.50 percent higher annual dividend yield, while matching the characteristics of the S&P 500 on historical beta and economic sector exposures.

How did the Yield-Tilt Fund perform? Initially not well. In 1978 unanticipated large increases in interest rates negatively affected the performance of high-yield stocks. Stock prices had to fall for income-stock yields to compete with the abnormally high yields offered in the fixed-income market. It was not until several years later that the performance of the Yield-Tilt Fund made up the ground lost in its first year. But it did recover.

Several lessons were learned from establishing and managing the Yield-Tilt Fund. It is important to be aware that common-factor performance often correlates with changes in macroeconomic variables, such as real GNP growth, and unanticipated changes in interest rates. Communication with the client before the fact of the macroeconomic risks is very beneficial in maintaining an amicable relationship in adverse economic environments. Fortunately, the client stuck with the Yield-Tilt Fund. Unless further increases in interest rates were expected, there was no reason to abandon the strategy. The Dividend Discount Model continued to show high-yield stocks as attractive. When interest rates eventually fell, the fund made up the ground it had lost. The yield-tilt experience also demonstrated the importance of having a sound set of reasons for engaging in a common-factor investment strategy, including an updated assessment supporting the continued implementation of the strategy.

Common-factor investing requires considerable judgment in the selection of target benchmarks, identification of common factors, and the development of portfolio revision and trading strategies. In the past decade, the proliferation of common-factor data bases and the relatively simple and low-cost computer systems required to manage common-factor-tilt strategies have opened up the opportunity for large and small investment management organizations to offer common-factor-tilted products.

There has not been a big rush by investment managers to offer such products. Conventional managers are in general disdainful of computerized investment processes. Some view it as trivializing investment management. Others simply are concerned that lower fees are generally charged for common-factor-tilt management, and its employment will hasten the collapse of the large profit margins the industry has historically enjoyed.

It also appears that a number of the quantitatively oriented people who are naturally drawn to indexing are petrified at the prospect of having to make investment judgments. If the task is to set up and manage an S&P 500 Index Fund, great! Don't ask them for an economic analysis, and don't expect them to accept unconditional uncertainty. Investment management without any judgment appears to be the objective of far too many "quants." This is the likely explanation for the slow development of common-factor-tilt strategies by organizations that have been well positioned to offer them.

Valuation Models

According to the Capital Asset Pricing Model, investors should price stocks such that their expected return falls on a security market line. Just how investors formulate their return expectations is not explained by CAPM. In 1971, Bill Fouse and Bill Jahnke, with the cooperation of Wells Fargo's Security Analysis Group, produced estimates of expected return by employing a version of the Dividend Discount Model (DDM), first suggested by J. B. Williams in 1934.[8] The security analysts provided 25 years of earnings and dividend forecasts, and the computer calculated the internal rate of return for each stock that discounted the forecast stream of dividends, plus a common terminal dividend growth rate for each stock, back to current price. The internal rate of return calculated from this process was annualized and was labeled the *expected return.* If, by chance, the projection of future dividends were accurate, the actual rate of return to be earned by investing in the stock in the long run would equal the expected return. It was recognized that the determinant of the expected return was the implicit expectation of the return on equity capital and that dividend policy was governed by how relatively successful the company is likely to be in investing retained earnings.

To determine if a given stock is an attractive investment relative to other stocks, it is necessary to consider differences in risk. The higher the risk, the higher the expected return should be. Sharpe had pointed out that only systematic risk difference should matter to investors, because nonsystematic risk is easily diversified away. To risk adjust expected returns, they were regressed on historical beta calculations. The result was an estimate of the ex ante security market line. The fitting of the security market line in June of 1970 was upward sloping, the average expected return was 11 percent, with an zero beta intercept of 9.4 percent. Ninety percent of the 270 stocks had expected returns positioned within 200 basis points of the line.

The natural question was whether or not the stocks plotting above the line would turn out to be good investments. To answer this question, two measurements of relative value were produced. The first measurement was to subtract from the expected return for each stock the equilibrium expected return for

[8] Jonn Burr Williams, *The Theory of Investment Value,* Cambridge: Harvard University Press, 1938.

stocks of similar risk taken from the security market line. This difference was called *mispricing*. Mispricing represents the deviation of the stock's expected return from the equilibrium expected return taken from the security market line, given the stock's beta. Mispricing was reported as an annual rate of return to be earned in perpetuity. The second measure of relative attractiveness was to determine the fair-market value of the stock. This was done by discounting the stream of dividends by the equilibrium discount rate taken from the security market line, given the stock's beta. The estimate of fair value was divided by the current price to produce the *present value ratio*.

The performance of these two measures of relative value were tested by periodically sorting stocks by their mispricing and present-value ratios into quintiles and measuring the periodic returns. This process produced large differences in performance between the highest and lowest rated quintiles on the order of 15 percentage points per annum through the 1970s. Several interesting relationships emerged from analyzing the characteristics of mispriced stocks. Low p/e stocks, high-yield stocks, and smaller-capitalization stocks generally showed up as attractive investments. Higher-beta stocks offered higher expected returns before risk adjustment, and the returns to beta were lower than forecast by original CAPM but consistent with CRSP research. Stocks with low market liquidity also showed up as offering superior return prospects. One interesting observation was that the popular growth stocks of the day, the so called nifty 50, were positioned well below the security market line, prophetically implying that a period of poor performance lay ahead.

It would take several years of paper portfolio testing and evaluation before management would accept the Wells Fargo DDM as the governing discipline in stock selection. The reason for the eventual acceptance of the model was relative performance. The stock selections of the Security Research Group were evaluated side by side with the selections of model. It was no contest. The model handily outperformed the analysts' stock selections. One analyst was so angered by the results of the performance derby, as it was called, he wrote a memorandum to management charging "the measurement of security-analyst performance is the cause of dysfunctional behavior" and requesting that all such measurement should cease.

Although one could criticize the Wells Fargo DDM on grounds of the rigor in the preparation of dividend projections and the reliance on historical beta, the process worked remarkably well at picking stocks. Additional tests of the model were conducted on five European countries and Japan by Sam Campopiano in the mid 1970s. Although intrigued but mostly incredulous, major foreign brokers and banks provided the necessary inputs to the model. Over an 18-month test period, the model demonstrated an ability to pick stocks in every market, even in Japan. The press of new business at the bank halted the experiment, since the opportunity to manage portfolios consumed existing resources.

The administration of the DDM presented a major problem. Analysts had a tendency to believe that good stocks could maintain above-average growth rates for years and even decades. These overly optimistic expectations were reflected in their forecasts. Empirical analysis of earnings growth rates showed that, after adjusting for dividend pay out ratios, earnings growth rates took a random walk. Past growth rates did not forecast future growth rates. Analyst forecasts of future growth faired only slightly better. Analysts had some limited ability to forecast earnings growth for several years into the future, but after the third year, the light in their crystal ball all but disappeared. It is surprising that Wells Fargo DDM performed as well as it did, given its reliance on analysts for estimating the duration of above-average growth for high-expectation stocks. Although there is no way of knowing, it is interesting to speculate on how much weight analysts placed on future growth rates in determining the appropriate p/e ratio in their traditional approach to stock valuation. Needless to say it was too much, since their stock rankings underperformed the rankings from the Dividend Discount Model employing their forecasts.

The common belief of sustainability of growth rates by good companies also explains the success of low p/e investing in recent decades. The failure of growth stocks on average to live up to investor expectations over time has resulted in a four percentage point per annum performance advantage for low p/e stocks relative to the market. As long as the market continues to pay too much for the illusion of extended superior growth, the rewards to low p/e investing can be expected to continue. A look at the distribution of p/e's provides a quick and easy check on the likelihood of continued success in low p/e investing.

There were other problems in the administration of the DDM. It was difficult to get analysts to produce forecasts consistent with a common set of macroeconomic expectations. Dealing with individual analyst tendencies for optimism or pessimism, getting analysts to make timely adjustments to their estimates in light of unanticipated major macroeconomic events, and controlling for game playing on the part of the analyst who wanted his favorite stocks to come out as looking attractive were all difficult to control.

Several refinements were made to the Wells Fargo stock valuation model in the mid 1970s. It was discovered that high-yield stocks tended to plot above the security market line. It was suggested that tax effects could account for this. To adjust for the yield effect, expected returns were regressed against both beta and yield. This produced a security market plane. The idea was that stocks plotting above the plane were attractive investments apart from the systematic returns associated with beta and yield. The person behind the development of the security market plane was Bill Sharpe, who had become a consultant to WFIA after Rosenberg had left to form BARRA.

This empirical existence of the security market plane had important theoretical and investment implications. The idea that stocks were priced based on a single risk factor was brought into doubt along with the notion that the market portfolio is necessarily optimal. Further investigation exposed the tendency for small capitalization stocks to plot above the security market plane. A regression of expected return against beta, yield, and the log of market capitalization produced the security market hyperplane. Stocks plotting above the hyperplane were considered underpriced. The explanation for small stocks plotting above the security market plane was illiquidity and trading cost. To test the illiquidity thesis, Fouse regressed mispricing from the security market plane against liquidity measures and found that illiquid stocks offered higher expected returns than liquid stocks after adjusting for beta, yield, and size. What was fascinating about these studies was the emergence of a complex structure in the pricing of stocks. Fouse referred to this structure as the "market's pricing mechanism." Stock prices could be evaluated relative to this structure to determine relative attractiveness. The current and historical slope coefficients on the returns to the factors could be evaluated in terms of their strength in the pricing of stocks. Increases or decreases in the slope coefficients were coincident with the movements of classes of stocks. Increases in the slope coefficient for market capitalization was associated with a poor period of performance for small capitalization companies. It was like looking inside a clock—price movements the hands, the market mechanism the gears and spring.

In 1981 Charles Pohl researched for Wells Fargo the relationship between stock performance and revisions of earnings expectations by Wall Street analysts. Using Wall Street consensus earnings revision data supplied by Lynch, Jones, and Ryan, Pohl discovered that stock prices had not fully adjusted to the revisions prior to the date the information was distributed by Lynch, Jones, and Ryan to its' clients. In fact subsequent price adjustments continued to occur over a five-month period following the publication of the consensus forecast earnings revision. Ongoing study of earnings revision indicates that it has continued to provide tradable investment information through the 1980s.

Still other academic studies in the 1980s on technical and seasonal factors indicate that numerous stock-market inefficiencies have existed, which individually or in combination with the fundamental and earnings revision data would have produced superior investment results. In 1988 Bruce Jacobs and Kenneth Levy published a paper in the (May/June) *Financial Analyst Journal* titled "Disentangling Equity Return Regularities: New Insights and Investment Opportunities." In the paper, the pure historical returns to 20 common factors, including earnings revision, were reported to have been large and statistically significant. The Jacobs and Levy paper, for which they received a Graham and Dodd award from the Financial Analyst Federation, is representative of some of the econometric research seeking to identify mispriced stocks through their common factors exposures. Given stock evaluations from such multifactor models, structured portfolio-management programs can be devised that control for desired portfolio exposures.

But there are those who challenge the idea of managing portfolios based on what they consider to be casual empiricism. Run enough regressions of stock prices against enough common factors and you are bound to find something that was produced by chance. And if you do find something that represented a true pricing anomaly, chances are that enough investors know about it that it no longer exists. To this point Fouse writes in 1974: "In the absence of a sound theoretical framework, one that passes the twin tests of logic and empirical verification, how can we plan successfully or manage prudently? In other words, without a valid model, the mere mechanical manipulation of the numbers in a problem does not necessarily make sense just because you are using arithmetic." Interestingly, a number of the common factors identified as producing superior performance were the factors that could be seen to promise superior returns before the fact, according to Fouse's market mechanism.

Portfolio Optimization

Portfolio optimization is the process of finding the investment portfolio that best meets explicit portfolio objectives. Portfolio optimizers are computer programs used to determine the best set of purchases and sales that bring the portfolio in line with its stated objectives. Generally speaking, investment managers have been slow to take advantage of portfolio optimization in the management of portfolio strategies designed to beat the market. In many cases the organizations did not have the in-house expertise to build the software, and the software available from outside vendors was prohibitively expensive. Besides, most portfolio managers did not want a computer terminal on their desk. As one newly minted Wharton MBA in the late 1970s observed when refusing the opportunity to have a terminal installed in his office, "upwardly mobile professionals wouldn't be caught dead with a computer terminal." Computers were rightfully seen by portfolio managers as devices that could control what stocks they purchase and sell. It really did not matter whether the stock selections came from a computer model or from an investment committee. While some of the managements of investment-management firms were inclined to want to try to use the computer to better control the investment process, most of those were not willing to fight the political battle with the portfolio managers.

In some cases portfolio optimizers were given a fair evaluation and found to produce unacceptable trade programs. This was sometimes the case when portfolio optimization used a covariance matrix drawn from past returns that did not reflect managers' expectations. Modern Portfolio Theory operates with the assumption that the portfolio manager agrees with the covariances contained in the covariance matrix. Where the manager disagrees, the solution is to construct a covariance matrix with which the portfolio manager can agree or to use a portfolio optimizer that does not require one.

The innovation in the use of computers to manage active investment strategies has come for the most part from new entrants in the business without large established interests. This is true even within the established organizations that have permitted quantitative investment management to develop and coexist in their organizations. The more successful applications of portfolio optimization tend to use the computer as a tool to control portfolio exposure to stock valuation models for reward enhancement and common factors for diversification. Portfolio managers often feel more comfortable explicitly controlling industry and common factor exposures using systems of targets and penalties. Unacceptable solutions are dealt with by resetting targets or penalties on the offending industry or common factor. Well-designed programs can take a portfolio manager through a complete cycle of target and penalty setting, optimization, and optimization review in several minutes. Generally within several cycles, a portfolio optimizer should produce an acceptable trade list. Although not employed in the optimization routine, a fundamental risk model is used to provide an unbiased estimate of specific risk and total risk. The estimates of risk are thought to be unbiased, because the stocks selections made for the portfolio are done independently of the covariance matrix.

Portfolio optimization programs should be generalized to optimize on any index, normal portfolio, or benchmark. The user should be able to choose the universe of eligible investments, maximum absolute and relative holding size and purchase weights, freeze stocks in or out of the portfolio, choose to or not to control directly on specific risk, select and weight valuation models, and, by exception, target industries and common factors. The portfolio optimization program should be accessible from the managers desk.

The decision to use portfolio optimization to manage portfolios, once a defined process for identifying the relative attractiveness of stocks has been established, is largely a matter of efficiency. A portfolio optimizer can generate portfolio solutions usually much faster than a portfolio manager can. This is especially true as the number of objectives for the portfolio increases. The computer will also tirelessly monitor and review portfolio holdings and investment opportunities—something portfolio managers are unable to do. It makes sense to let the computer do the data processing while the portfolio manager oversees the portfolio revision process, spending more time in research, trading, and client service.

The growth in power of the consulting community in the hiring and firing of investment managers has and will increasingly promote the use of portfolio optimization programs for defining normal portfolios to serve as manager-specific performance benchmarks and tools for controlled implementation of predefined investment strategies. Consultants are interested in defined investment styles that are subject to analysis and control in a multiple-manager framework. It is important to the consultant that the manager can and will operate within the role that has been agreed upon. Managers who can demonstrate that they have well-defined and controlled investment processes have an advantage over managers who can not in the absence of material differences in past performance.

Asset Allocation

One of the major successes in the structured investment management business has occurred in the area of asset allocation. Asset allocation seeks to define the optimal exposure to asset classes as defined by the investment manager. Typically stocks, bonds, and Treasury bills are selected as asset classes, and an asset-allocation model is used to determine the optimal portfolio allocations. More esoteric asset allocations across international equity markets, active managers, and industrial sectors have been determined using asset allocation models. Recently programs that look at the optimal asset allocation, taking liabilities into consideration, have been developed.

When Bill Fouse arrived at Wells Fargo in 1970, he found a fertile ground for developing quantitatively oriented investment management products. One of the first products Fouse and his Investment Systems Group tackled was to develop an Asset-Allocation Model, which was spiritually in tune with the work of Markowitz. Fouse proposed to determine the appropriate mix of stocks and bonds in a portfolio based on a utility-maximization routine that made some assumptions about pension-plan risk, taking preferences and the distribution of expected returns to a selected investment horizon. The distribution of expected returns assumed that the annual returns used in the model were log-normally distributed. The mean expected returns for stocks were taken from the Dividend Discount Model, and the expected returns for bonds were taken from the yield curve for the investment horizon. The post World War II period was chosen to use for standard deviation and correlation estimates. Barr Rosenberg provided a formula for converting the annual return and risk estimates into a holding-period return distribution. By numerically passing the expected return distribution by a numerical representation of pension-plan risk preferences, the utility of any particular mix of stocks and bonds could be determined. By evaluating different portfolio mixes of stocks, the one offering the highest utility could be found.

In the 1970s it was fairly common for an investment manager to determine the appropriate mix between stocks, bonds, and cash in a portfolio. It was also recognized that the decision was an important one, perhaps the most important investment decision to be made. Investment managers would adjust asset mixes based on their feel for the market. Quite often how they felt was governed by how the market had recently done. Investment managers liked stocks in up markets, and liked cash in down markets, and generally found bonds to be boring. The problem was that investment managers defined *up* and *down* markets by looking backward and acting as if the trend would continue. Not surprisingly, the balanced account results did not look very good, especially against a benchmark of 60 percent stocks and 40 percent bonds, which someone had discovered in mining the CRSP data base to have provided a pleasing trail of risk/reward tradeoffs and which also seemed to be fiduciarily sound. One reason the 60/40 mix tended to do well was that the benchmark was rebalanced yearly. If stocks did well in a given year, some of them would be sold to rebalance back to 60/40. If stocks did poorly, they would be purchased at the next rebalancing. This was a good thing, because good returns tended to follow bad returns and visa versa, or, as the academics pointed out, "the markets tend to exhibit mild bouts of negative auto serial correlation." Investment managers were operating under the rule "cut your losses and let your profits run," and losing big, according to Larry Tint, who ran the Pension Fund Advisory Service at Wells and who worked for the Merrill Lynch Performance Measurement Service in the early 1970s.

The Wells Fargo Asset Allocation Model was not an instant marketing success. In the early 1970s, nobody was managing balanced accounts using a computer model. The performance of the model was evaluated by real time on paper, and in 1977 the strategy was implemented using funds provided by the State of Delaware Public Employee's Retirement Fund. The S&P 500 Index Fund was the logical investment vehicle to implement the swings into and out of stocks. In order to eliminate the transaction cost impact to investors in the index fund not party to a contribution or withdrawal, a method was devised in 1973 for charging transaction costs back to the parties entering or exiting the Index Fund. Some years later, a competitor learning about the practice "blew the whistle," forcing Wells Fargo's Trust Counsel, John Barnes to obtain an exemption from the Comptroller of the Currency to continue the practice of allocating trading costs to fund participants.

And how did the Asset-Allocation Model work in practice? Beyond expectations! The five-year returns for the Asset-Allocation Model were usually above the fifth percentile for balanced accounts, as reported by the SEI fund evaluation service. The original concept was to determine the strategic asset mix to a long-term investment horizon based on long-term expectations for risk and return. There was no promise that the model would perform well in the short run. But the model did perform well. What happened was that a decline in prices for an asset class would, in the absence of adjusting the cash flows, dividends or interest, bid up the expected return. The model would then call for a higher commitment to the asset class. Sounds like "buy low and sell high." Just the opposite of what the portfolio insurers were doing. As long as prices fluctuated around some secular equilibrium and the particular settings for the asset allocation model were not so severe as to produce a 100 percent allocation to one asset class all the time, good performance followed. Marketing euphoria and emulation by other organizations followed. What was *strategic* asset allocation was now called *tactical* asset allocation. Something was lost in the stampede.

When the original asset-allocation model was set up at Wells Fargo, programs were devised that would define holding-period returns for changes in the structure of capitalization rates, discount rates for stocks, and yields for bonds. Short-term holding-period returns could be calculated based on any capitalization rate structure scenario. Short-term return expectations could be input to the asset allocation model. The problem facing tactical asset allocators is whether the recent price changes reflect a secular movement towards a new equilibrium structure of capitalization rates or the underlying expectations for future cash flows have shifted. Enough to keep any serious tactical asset allocator up at night!

Judgment

Structured portfolio management employs computer-based investment technologies to manage investment strategies that are consistent with investment theory and empirical evidence. The early applications usually involved little in the way of traditional investment judgments. Most structured investment managers historically have attempted to minimize judgment in their investment processes. Developments that affect human perceptions for which there is no objective empirical evidence, computer-based or otherwise, are avoided. Should such perceptions be incorporated? How far can structured portfolio managers go down the road of employing forecasts and judgments before they lose their legitimacy? Tough questions!

The hardline structuralists have and continue to voice their concerns about *subjectivity* in the investment-management process. The fewer the judgments, the better. For them, market portfolios are the answer. Others perceive an important role for *organized subjectivity* in the investment management process. James Vertin presented the opportunity to his organization on the eve of structured investment management in 1970:

> The largest challenge facing us, now and tomorrow, is posed by the problem involved in integrating our human insights and unique forecasting ability into what is, in its present form, a management method which bypasses such inputs. Were there no Management Sciences Department skilled professionals exercising their proper functions will provide added value to the results of any asset management system. We must so structure our organization as to produce these judgments in the form and content needed. By starting now to bring the theory to bear to the fullest extent possible within the framework of our activities while

> simultaneously utilizing management science skills and the computer in their implementive context, we will be in a leadership position relative to what will eventually develop as the mainstream of evolution in the asset management business.

Some of the earlier academic beliefs regarding the efficiency of the stock market and the futility of successfully employing technical and fundamental analysis have been shaken by studies documenting the existence of stock-market anomalies. Yet the record of conventional market timers and active equity managers fails to provide evidence that traditional practice adds value. The results of organizations employing quantitative methods to produce value-added results have produced some startling successes, as well as some failures. The gradual incorporation of quantitative methods by traditional investment managers and the sale of higher-margin, value-added products by indexers has in some cases blurred some of the distinctions between the two.

The ultimate choice of whether Vertin's 1970 vision of "integrating human judgment" represents the "mainstream of the evolution in the asset management business" in the 1990s and beyond, or a side game played in the future at the indulgence of a decreasing minority of pension plan sponsors, who continue to search for what Paul Samuelson describes as "needles that are so very small in haystacks that are so very large," lies in the hiring practices of the corporate and public pension-plan sponsors. Whether integrating human judgment in either case necessarily represents what Emerson refers to as "flights of fancy" is of central importance to all interested parties, investment managers, pension fund sponsors, and pension beneficiaries. Certainly the burden of proof is on those who claim the ability to integrate judgment. As the information age in which we live continues to develop, it will become increasingly more difficult to avoid agreeing with the unknown author's sage observation, "none of us is as smart as all of us," at least when it comes to setting security prices.

CHAPTER 13

Enhancing Portfolio Returns with an Active Currency Overlay

Philip Green
Bankers Trust Company

Introduction

This chapter begins with a short discussion of market efficiency and the foreign exchange market. Market efficiency is an important feature of any market, since the presence or absence of efficiency will determine which investment strategies will add value. Given various degrees of market efficiency, we then turn our attention to identifying those investors and investment strategies that have the ability to beat the foreign exchange market. The final section outlines an investment management framework where an optimal (in risk/return terms) currency portfolio is constructed. It is shown that a currency portfolio that holds currencies that are "perceived" to have relatively high risk will add incremental value regardless of the degree of market efficiency.

Market Efficiency

Market efficiency can be segmented into three degrees: weak form, semi-strong form, and strong form. A weak-form efficient market is one in which all information contained in historical prices is reflected in current prices. A semi-strong form efficient market incorporates all publicly available information in its prices, while prices in a strong-form efficient market reflect all information, both public or private. Strong-form efficiency implies the existence of semi-strong form as well; likewise, semi-strong form efficiency implies weak-form efficiency.

In the absence of strong-form efficiency, investors can expect to beat the market only if they possess either better information than the market or a superior information processor or model. If the market is strong-form efficient, though, investors have no alternative but to endure more systematic risk than the market.

Much empirical work has focused on market efficiency and the foreign exchange market. However, the results are inconclusive because statistical tests of market efficiency tend to be structurally weak. In particular, standard tests actually represent joint tests of both market efficiency and the model employed to estimate the market's expected return. As a result, even if the foreign exchange were truly efficient, we could still reject market efficiency if our expected return model is inaccurate. Unless we are confident in our expected return generating model, we cannot render a definitive verdict on market efficiency based upon standard empirical tests alone. To date, many market efficiency tests have used the forward rate as the market's expectation of future spot. However, recent works on the foreign exchange market (Fama 1984; Koedijk & Ott 1987) have cast doubt on whether observed forward rates indeed reflect the market's expectation of future spot. These studies suggest the existence of risk premia imbedded in forward rates. The ambiguous results of empirical tests suggest we should proceed with caution when making judgments about market efficiency based upon statistical studies alone.

On the other hand, probably the strongest evidence in favor of foreign-exchange market efficiency is the current state of affairs in the foreign exchange market. In particular, the enormous size of the foreign exchange market, as well as the fact that armies of highly trained professionals constantly scour the currency markets for investment opportunities, means that any perceived inefficiency will be arbitraged almost immediately.

A popular argument against market efficiency is the existence of nonprofit-motivated investors, namely the central banks. However, even if central banks are willing to lose in order to achieve a perceived greater good, their losses would not necessarily be evenly distributed across all other market participants. Those participants who have an informational edge will reap a disproportionate amount of any spoils generated by central bank activity. Consequently, central bank activity does not preclude the market from being semistrong-form efficient.

Anatomy of a Winner

Which market participants can we expect to beat the foreign exchange market? In the absence of strong-form efficiency, the winners will be those players who have either superior models or superior information. Given that models are rarely known to lose money in historical simulations, we need a methodology to distinguish between a multitude of high performance models. The simplest answer is to pose three questions:

- Why did the model work in the first place?
- Why should someone believe that these same conditions will continue in the future without anyone else noticing them?
- What incentive does the owner have in sharing the model with others, since that would merely reduce his own potential profit?

Most models are offspring of one of two genres: time series or structural. Time-series models examine historical price movements in an effort to predict future prices. As such, the success of time-series (a.k.a. technical) models depends on the absence of weak-form efficiency. Structural models attempt to forecast future prices based upon historical relationships of fundamental, usually macroeconomic, variables and current information. Similarly, the success of structural models depends on the absence of semistrong-form efficiency. A number of studies have shown that both time-series and structural models have a poor track record of predicting future foreign exchange prices (Meese & Rogoff 1983), implying the existence of semi-strong form efficiency.

Arguably, the players who have the greatest edge in terms of information are foreign exchange dealers. Not only do they employ scores of highly trained individuals analyzing all available information, but they also possess knowledge of trade flows. Trade flow information is extremely valuable for many reasons. Trade flows reflect information about market sentiment. Therefore, dealers can position themselves in order to capitalize on shifts in market sentiment before other participants. In addition, trade flows reflect investors' strategies and, therefore, future intentions. For example, if dealers can infer investment strategy from trade flows, such as hedging the currency exposure of a foreign denominated asset, they will know that this investor will return to the market to repurchase his short foreign currency position either in the form of a swap or spot transaction. Finally, trade flows signal government policy. Early knowledge of central bank activity allows dealers to position themselves ahead of the rest of the market. In order to identify particular players claiming to have superior information, though, we need a formal evaluation procedure.

Identifying the Winners

How can we distinguish between a dealer/manager with superior information and one who reads *People* magazine? Actually, it is difficult on an ex-ante basis.

Unlike, perhaps, baseball, past performance in currency management does not necessarily imply future success. Fortunately, on an ex-post basis, we can use a yardstick proposed by Merton (Merton 1981). Merton developed his measurement system to identify those managers who claim to have skill at timing risky asset markets such as the foreign exchange market. By market timing, we mean forecasting movements in exchange rates (i.e., the dollar will appreciate against the yen). Extending the baseball analogy, Merton suggests separating a manager's tract record (batting average) into two subcomponents: namely, the manager's performance in strong-dollar environments versus his performance in weak-dollar environments. Each subcomponent record would be weighted equally in evaluating the skill of a manager. In such a framework, scores will range from a high of 2 (attained when a manager predicts 100 percent of the upmarkets correctly as well as 100 percent of the downmarkets) and a low of 0 (predicts every market incorrectly). A manager with no skill will fall between 0.9 and 1.1.

To understand the significance of equal weighting, we continue with the baseball analogy. In baseball, a player who bats .500 against righties but .000 against lefties would have a higher batting average than a player who bats .300 against both as long as the percentage of righties in the league exceeded 60 percent. However, if the probability of facing righties versus lefties in the future is not correlated with historical averages, then a weighting scheme based upon future expectations is appropriate. If currency markets are weak-form efficient, future movements in foreign exchange rates will not be correlated with past movements, and the probability of a strong-dollar environment versus a weak-dollar environment is 50/50. As a result, currency managers should have their historical component records equally weighted, since an equal weighting scheme reflects future expectations about the foreign exchange market.

Further Implications of Market Efficiency

Within an efficient or rational market, investors are compensated for enduring systematic risk. For example, an investor holding a diversified portfolio of equities will expect to receive greater compensation than an investor who holds U.S. Treasury bills, because the market perceives equities as the riskier investment. The same notion holds in the foreign exchange market. In particular, if the market perceives the Australian Dollar (AUD) to have more systematic risk than the West German Deutschemark (DEM), then a long AUD, short DEM position would have a positive expected return. If it did not, then no rational investor would hold it long; rather, they would want to short it. Hence, in order for an equilibrium condition to exist, the AUD/DEM position would have to be priced so that long investors would have a positive expected return while short investors would have a negative expected return.

How can we be sure that the foreign exchange market perceives currencies to have unequal risk? One indication is to compare the implied volatilities imbedded in currency option prices. If the market perceived every currency as equally risky, implied volatilities should be similar. In practice, they are not.[1]

A second indication that all currencies are not perceived to be equally risky is the fact that forward rates have consistently misestimated future spot rates.[2] In order for forward forecast errors to occur consistently over time, either the foreign exchange market has to be inefficient or risk premiums must be imbedded in forward rates so that investors who hold currencies with higher systematic risk are duly compensated.[3] In other words, if the market perceived all currencies to be equally risky (and the market is efficient), then the expected return streams arising from ITL (Italian Lira) or CHF (Swiss Franc) investments would be the same and, on average, this return stream should be close to zero over a significantly long time period. We shall see that this has not been true since 1978. The implication is that the market does not perceive all currencies to be equally risky and currency risk premiums exist, imbedded in forward rates.

Constructing an Optimal Currency Portfolio

How can we capitalize on the fact that different currencies have different levels of perceived risk? In order to construct a currency portfolio that maximizes expected return at any level of risk, we first need to quantify the risk and return attributes of every currency. The expected return of each currency reflects an estimate of that currency's risk premium. Two general approaches exist for estimating the size of the risk premium. The first approach would be to build a currency risk premium model that would involve multiple variables and complex relationships (such as lagged influences). Though theoretically appealing, the construction of a risk premium model has proven to be extremely difficult (Bomhoff & Koedijk 1988; Hansen & Hodrick 1983).

An alternative methodology is to find indicator variables that are highly correlated with currency risk premia. We have identified two market variables that we believe to be significant predictors of currency risk premiums: the global term structure and information imbedded in currency option prices. The global

[1] Caveat: Note that since implied volatility measures total risk and not systematic risk, we need to be careful to compare currencies that have similar correlations with the market portfolio. For instance, EMS currencies have similar correlations with other assets because they have such high correlations with each other.

[2] Cumby presents evidence of statistically significant ex-ante returns to forward speculation, but concludes that consumption-based models of risk premia fail to provide an adequate description of these returns (Cumby 1989).

[3] Actually, other explanations do exist. Karen Lewis has suggested that forward rate forecast errors may be due to a change in the process of a fundamental variable that is not immediately recognized by the market. The market then requires time to learn about the new process (Lewis 1989).

term structure represents term structures across all capital markets. The intuition behind using nominal interest rates is that nominal rates reflect expectations about both real interest rates and inflation.[4] Therefore, differences in nominal rates reflect differences in expectations about real rates and inflation. Currencies that have high nominal rates have either higher real rates of return or higher inflationary expectations than other currencies. However, in either case, investors who hold the higher interest rate currency receive added compensation (in the form of a greater real rate or more compensation against inflation) than investors in lower interest rate currencies.

The rationale behind using option prices is that option prices convey better information about risk and return than the cash market. As long as investors with superior information place their bets in the options market (in order to gain the maximum leverage available), then the information set imbedded in options prices will be superior to the information set imbedded in cash prices (Peterson & Tucker 1988). Various statistical methods allow us to extract both risk and return estimates from options prices.

Empirical Results

Our research strongly indicates that, historically, the market has not treated forward positions in the world's major currencies as perfect substitutes, and that information is available, ex ante, regarding the market's perception of currency-specific risk.

Table 1 contains the annual cash flow that would have accrued to an investor who sold forward a currency that appeared, ex ante, to be marginally "less risky," and simultaneously purchased forward an equal amount of another currency that appeared marginally "more risky." In our study, 30-day positions were taken, and profits and losses were realized on a buy-hold basis. Transactions were executed at the prevalent bid-asked prices. Profits were reinvested, and losses were financed for 30 days at the prevailing Eurodollar rate. Cash flows are reported as a percentage of the total long (or short) forward position.

Two strategies are reported. "Free Float" refers to positions taken from a universe of freely floating exchange rates. "EMS" refers to positions taken from a subset of the currencies within the EMS that are highly correlated. Notice that in both cases, the average cash flow is not zero—that is, over the 10-year period, the market did not perceive forward positions in these currencies to be perfectly substitutable. The hypothesis suggesting the existence of currency risk premia is strongly supported by this data.

[4] The well-known Fischer equation postulates that nominal interest rates are equal to a real rate plus an expectation of future inflation.

Table 1 Historical Currency Risk Premia Earned

Year	EMS Portfolio (%)	Free Float Portfolio (%)
1979	6.3	18.6
1980	16.4	28.1
1981	4.3	10.2
1982	4.1	16.8
1983	8.0	(0.4)
1984	8.6	11.8
1985	4.1	13.3
1986	9.7	(14.0)
1987	1.3	9.2
1988	9.2	20.6
Total (annualized)	7.2	11.4
Standard Deviation	4.4	9.9

Note: Risk premia refers to realized cash flow divided by total long (or short) forward position + lending proceeds – borrowing costs.

In Table 2, the risk premia earned by the two currency positions are compared with four examples of risk premia that are "universally accepted" in the field of investment finance: domestic equities versus Treasury bills; small-cap stocks versus large-cap stocks; long-maturity government bonds versus Treasury bills; and corporate bonds versus government bonds. Notice that over this period the existence of all risk premia, with the exception of currency, would be rejected at the 95 percent significance level. Moreover, the existence of currency risk premia is statistically significant to a degree of confidence exceeding 99 percent.

Conclusion

The historical evidence is clear: the market has exhibited distinct assessments of marginal risk to forward positions in various foreign currencies. Assuming investors are rational and markets are efficient, expected returns to forward positions in various foreign currencies are not zero but are commensurate to their market perceived uncertainties. Consequently, rational investors embarking on an asset allocation policy, including potential investments in foreign assets, should manage the exposure to currencies in the same manner as that done for other asset candidates. Moreover, because the exhibit distinct risk and return characteristics, forward exposure to foreign currencies should be considered on an individual basis and not as a single asset. In addition, attempting to market-time the

Table 2 Significance and Comparison

	Equity Premium	Small Stock Premium	Maturity Premium	Default Premium	Free Float Currency Premium	EMS Currency Premium
No. of observations	120	120	120	120	120	120
Minimum	–21.98	–9.77	–9.25	–4.77	–8.09	–4.04
Maximum	12.96	9.27	13.81	4.06	8.59	4.94
Sample mean	0.65	0.23	0.20	0.03	0.95	0.60
Sample Std Dev	4.76	3.14	4.12	1.29	2.86	1.26
Std Error	0.43	0.29	0.38	0.12	0.26	0.12
T-statistic	1.50	0.81	0.53	0.23	3.65	5.22
Significance	.86	.58	*	*	.99	.99

Other Premia Definitions:

Equity premium = S&P 500 return over Treasury bills.
Small stock premium = Small capitalization stock return over S&P 500.
Maturity premium = Government bond return over Treasury bills.
Default premium = Corporate bond return over government bonds.

* Indicates significance less than .50.

Source: Ibbotson Associates.

foreign exchange market by betting on the levels of particular exchange rates is a game fraught with hazards. Under semi-strong form efficiency, only participants with superior information will beat the market, and even then, identifying those participants with superior information is difficult.

References

Bomhoff, Eduard and Kees Koedijk. "Bilateral Exchange Rates and Risk Premia," *Journal of International Money and Finance*, June 1988, pp. 205–20.

Cumby, Robert. "Is It Risk? Explaining Deviations from Uncovered Interest Parity," NBER *Working Paper No. 2380.*

Fama, Eugene. "Forward and Spot Exchange Rates," *Journal of Monetary Economics*, November 1984, pp. 319–38.

Hansen, Lars Peter and Robert Hodrick. "Risk Averse Speculation in the Forward Exchange Market: An Econometric Analysis of Linear Models," *Exchange Rates and International Macroeconomics*, ed. Jacob Frenkel, 1983.

Koedijk and Mack Ott. "Risk Aversion, Efficient Markets and the Forward Exchange Rate," FRB *St. Louis Review*, December 1987, pp. 5–13.

Lewis, Karen. "Changing Beliefs About Fundamentals and Systematic Rational Forecast Errors: With Evidence From Foreign Exchange," Salomon Brothers Center *Working Paper No. 507.*

Meese, Richard and Kenneth Rogoff. "Empirical Exchange Rate Models of the Seventies: Are any fit to survive?" *Journal of International Economics*, February 1983, pp. 3–24.

Meese, Richard and Kenneth Rogoff. "The Out-of-Sample Failure of Empirical Exchange Rate Models: Sampling Error or Misspecification?" *Exchange Rates and International Macroeconomics*, ed. Jacob Frenkel, 1983.

Merton, Robert. "On Market Timing and Investment Performance: I. An Equilibrium Theory of Value for Market Forecasts," *Journal of Business*, 54, No. 3, 1981.

Peterson, David and Alan Tucker. "Implied Spot Rates as Predictors of Currency Returns: A Note," *Journal of Finance*, March 1988, pp. 247–58.

CHAPTER 14

Selecting a Global Custodian

Steven L. Fradkin
The Northern Trust Company

> We too must suffer all the suffering around us. What each of us possesses is not a body but a process of growth, and it conducts us through every pain in this form or that. Just as the child unfolds through all the stages of life to old age and death (and every stage seems unattainable to the previous one, whether in fear or in longing) so we unfold (not less deeply bound to humanity than to ourselves) through all of the sufferings of this world. In this process there is no place for justice, but no place either for dread of suffering or for the interpretation of suffering as merit.
>
> *Franz Kafka*

Franz Kafka's observation of human growth provides an unfortunate parallel to the experiences some investment professionals have had in evaluating and selecting a global custodian. The purpose of this chapter is to help plan sponsors, investment managers, and other professionals face these challenges by defining global custody and addressing some of the issues warranting review in the process of selecting a global custodian.

Global custody is a combination of bank services designed to facilitate the safekeeping, settlement, and reporting of a fund's worldwide securities transactions. It is a product of necessity. Many foreign markets do not permit the transfer of ownership certificates beyond their borders. This restriction coupled with the operational inconvenience and administrative expense of physically moving all securities purchased in foreign markets back to the United States for safekeeping has spawned the global custody industry.

The industry has not been without its faultfinders. Critics have challenged everything from reporting systems and personnel to the exchange trading infrastructure existing in some countries. Consider the following excerpts from *The Wall Street Journal*, which highlight just a few of the problem areas,

> Since Jan 22, (some) stock exchanges have halted trading at 2 PM, an hour and a half early, to keep down the volume of new paper work. And yesterday the "settlement period" during which investors and brokers must complete payment and delivery in a securities transaction was lengthened to five days from four.[1]
>
> On the average, the backlog is so large that it takes a transfer agent a week to process a change in ownership of a security, against two days a few years ago; thus, many transfers apparently still will become fails even under a five day rule.
>
> And as long as there are fails, back offices will be bedeviled by another problem: Dividends going to the wrong people.[2]
>
> It has become almost certain that all major securities exchanges . . . will close the next three Wednesdays and Friday . . . because high trading volume has swamped . . . clerical staffs and impaired the operating efficiency of scores of firms.[3]

From the perspective of a U.S.-based investor, accustomed to the highly automated and efficient depository environment existing in the United States, the custodial challenges associated with trading in some overseas markets can be understandably frustrating. It helps, however, to take a step back to garner perspective. The excerpts set forth above, while closely paralleling some of the difficulties global custodians face in overseas markets, are from the 1968 edition of *The Wall Street Journal* and are describing U.S., rather than foreign markets.

The point here is really twofold. First, global custodians have been down this road before. Major custodian banks have grappled with the fundamental issues associated with securities transaction processing. They have worked with other market participants to foment change in the United States and are, at least in some senses, somewhat better prepared the second time around to address settlement issues in overseas markets. Second, there is propensity to blame the problems associated with overseas settlement on the "less sophisticated or capable foreign markets." Evidence suggests to the contrary that, rather than being

[1] Richard E. Rustin, "How A Brokerage Firm Battles the Paper Work Trading Surge Creates," *The Wall Street Journal*, February 6, 1968, p. 1, col. 6.

[2] Richard E. Rustin, "How A Brokerage Firm Battles the Paper Work Trading Surge Creates," *The Wall Street Journal*, February 6, 1968, p. 1, col. 6.

[3] "Major Securities Markets Expected To Shut Next 3 Wednesdays, July 5," *The Wall Street Journal*, June 6, 1968, p. 3, col. 1.

less capable, many of these markets are simply at a different point on the developmental curve of exchange trading volume and consequent trading and settlement infrastructure. As interest and its associated trading volume continues to escalate on a sustained basis in these markets, the evidence suggests that, as in the United States in the 1960s, these markets will better develop their settlement and related processing infrastructure.

Global Custody Service Components

The primary components of a global custody service include:

- Safekeeping.
- Trade settlement.
- Foreign exchange.
- Cash management.
- Income collection.
- Tax withholding and reclamation.
- Corporate actions.
- Securities lending.
- Multicurrency monthly and on-line reporting.
- Account management.

There is no one right or best process for evaluating and selecting a global custodian. As with the selection of a new car, individual buyers attribute differing levels of importance to varying product features. There is no general agreement on the "most important" service component. Indeed, at The Northern Trust, we believe the most important element is the way in which a global custodian balances all of the service components to provide a timely and accurate service consistent with predetermined client objectives.

The Search Process

The first stage warranting consideration in preparing to evaluate prospective service providers is the development of a selection process or methodology. Some organizations develop the process in-house, while others engage a consultant to provide assistance A typical selection process might include the following steps:

1. Gather information about the fund and its related investment objectives.
2. Engage in preliminary research about global custody and develop a list of prospective service providers.
3. Develop a written request for proposal ("RFP") for prospective global custodians.
4. Develop a matrix to compare responses to the RFP.
5. Follow-up with investment manager references.
6. Follow-up with client references.
7. Prepare and submit follow-up questions as appropriate.
8. Select and visit finalists.

After the decision-making process has been established, the two most arduous tasks are developing the written RFP and comparing RFP responses.

The written RFP establishes the tone and focus of the search. It should be prepared with thought and should afford respondents an opportunity to clearly describe their service features and distinguish them from competitors'. The objective of a well drawn RFP is to elicit information. The best RFPs tend to be informational rather than confrontational in tone. Antagonizing prospective service providers can restrict the number of respondents, thereby limiting the amount of information gathered. RFPs that include open-ended questions may receive responses that are difficult to matrix but can provide valuable information.

A well-drawn RFP will elicit the finer technical points of providing a global custody service as well as the provider's organizational structure, experience, and philosophy of management. These latter issues should not be overlooked, as they are representative of a bank's commitment to the custody business and resulting support for future client needs.

In preparing an RFP, as with the search process generally, there are a variety of areas warranting consideration. You will want to draw the RFP to focus on the issues of greatest importance to your organization. To assist you in the development of an RFP, I have highlighted some of the issues you may wish to consider.

Background and Organization

Understanding a bank's history and the reasons for providing global custody services can provide important clues as to its long-term commitment and senior-level support for the product. A good RFP will force respondents to articulate their reasons for entering the business. A critical area of focus should be on the extent to which the provision of global custody services is an important component of a bank's overall strategy. Is the service simply part of a design to generate fee income, or does it relate more closely and logically to serving the bank's core clients? Banks that focus on custody services as a primary rather than peripheral business are more likely to allocate the resources—financial, systems, human, and other—necessary to maintain the highest quality service.

Examine the bank's experience in providing both domestic and global custody services. Commitment to the business should be measured by deeds rather than words. Look for evidence of commitment. This evidence can take many forms including staff additions, investments in technology, education, training, and other areas. The hallmark of a true commitment should be seen in consistency. The best providers tend to have demonstrated commitment to the custody business over time on a consistent basis.

Review the organizational structure of the global custody group within the context of the entire bank. Does the group have direct access to systems resources or must it lobby with the systems department? What is the experience level of the managers responsible for the product, and how deep is their staff? Has the bank experienced a great deal of turnover? The organization, tenure, and personnel of the group are important factors influencing long-term success. Try to elicit the bank's vision as to where the business is headed and what the respondent believes will place them at the forefront of the business in the future.

The number of clients, portfolios, and assets under administration is not always the best indicator of success. It does, however, provide a basic indication of credibility and evidence of the necessary scale to justify future investment and so should be considered. Remember that managing growth is every bit as important as growth itself. Be sure that the bank you choose is not growing so fast and without plans that it can not support the needs of its many clients.

Safekeeping Network

Be sure that your global custodian runs its own network rather than piggybacking off the network of another organization. Control of the network is essential in maintaining a quality service and being close enough to the real-world challenges to support client queries and concerns. By the same token, a bank having its own network is not sufficient of itself. Any bank can establish a network. The real issue is how the bank supports its network to make it more effective than those of its competitors.

A list of countries covered and agents and depositories used should be requested. The RFP should also explore the global custodian's appetite for establishing relationships in new markets if so requested and its process for so doing. Thorough and painstaking attention to detail at the inception of a foray into new markets is important and should be examined. Just as you will establish a thorough process for the review and selection of a global custodian, that custodian should, in turn, have a clear and detailed process for selecting and reviewing subcustodians.

The RFP should require respondents to carefully describe their operational and credit review process for subcustodian banks. How often is it done and by what areas of the bank? Has the bank recently made changes in any given markets and if so, why?

Trade Settlement

Trade settlement is an important component of a global custody service. Communication systems, from investment managers to the global custodian and subsequent communication to subcustodian banks and back to managers, should be efficient and unencumbered. The smooth flow of data in this highly information-intensive business plays an important role in the effectiveness and efficiency of a global custodian's network. Requesting a flowchart of how trades move through the system from trade receipt through settlement is a good starting point. Focus on the number of inputs upon the custodian's receipt of trade instructions. Remember that multiple inputs increase the probability of multiple errors.

Global custodians should also have a clear policy for follow-up and review of failed trades. Failed trades are an everyday part of the global custody business, so an internal tracking and action policy is important and should be examined. No custodian can guarantee that settlement will occur on time in all markets, but all custodians should have clear and sound policies for tracking and following-up in problem countries.

Foreign Exchange

The provision of foreign exchange services is a value-added service and should be evaluated as such. The global custodian should provide investment managers with a convenient variety of alternatives for handling foreign exchange, including acceptance of standing instructions and direct access to dealers to negotiate live rates.

The global custodian should have to earn foreign exchange business by providing competitive rates. Custodians should allow managers to execute foreign exchange outside their network and should be accommodating if a manager decides to pursue this route.

The ability to offer competitive block trading facilities is another area of coverage that may be included in the RFP.

Cash Management

Global cash management practices and capabilities may warrant particular scrutiny in the evaluation process. Investment of non-U.S. dollar cash balances can vary considerably among global custody providers. Country coverage should be extensive and rates should be competitive. Be sure to compare minimum investment amounts and time periods. The global custodian should also be able to provide longer-term, fixed-time deposits.

Income Collection

The income collection process in the United States and in foreign markets is relatively straightforward. Funds are credited on payable date in most major markets. An examination of exceptions to this rule and the custodian's average delay experience between payable date and receipt in problem countries is worth exploring. The ability to provide detailed reporting of income receivables should also be reviewed.

Tax Withholding and Reclamation

Many institutions are eligible for partial or total exemption from withholding taxes on foreign income. The custodian should arrange to have all income received with the correct amount of withholding tax deducted and to make reclamation of recoverable taxes in applicable jurisdictions. Organization of the tax reclamation group should be studied, since the nuances of operating in countries with differing tax laws and treaties requires the attention of a focused group to ensure maximum efficiency. Procedures for filing and tracking reclaims on a country-by-country basis should also be examined.

The reporting of tax reclaims receivable is another area of increasing importance among many global investors. A custodian's flexibility and ability to provide detailed reports of reclaim receivables may be another area warranting review.

Corporate Actions

Handling of corporate actions is an area of growing interest among fund sponsors and investment managers. The critical areas of exploration should include the custodian's information sources and notification process. The extent to which a global custodian is able and willing to forward proxy information to investment managers is another area you may want to explore in the RFP.

Securities Lending

Securities lending has become an increasingly important value-added service offered by some global custodians. Global lending is important for two reasons: it can provide additional incremental revenue to offset a portion of the custody fee; and, it provides evidence of a custodian's commitment to the overall business. While some funds to not envision participating in the lending program at the outset, a custodian's demonstrated commitment to the business and ability to meet the fund's future needs is important. An evaluation of a securities lending program should include a review of the risks, controls, countries covered, process, queuing, splits, and indemnification.

Reporting

Timely, accurate, and thorough reports are the watchwords of most groups evaluating global reporting capabilities. Reports should be multicurrency, fully accrued, and should provide details of gain/loss by both security and currency. Taxable entities should ensure their custodian can provide fully automated, multicurrency tax lot accounting and reporting.

Timeliness and accuracy of reporting are difficult criteria to evaluate. Each organization has its own time frames and should be cognizant of the relatively slower turnaround times all global custodians have at present. The extent to which the custodian provides full audits and on-line access should also be reviewed. The best basis for gauging timeliness and accuracy, however, is through client and investment manager references. Take full advantage of the ability to talk with a custodian's references.

Fees

A clear and rational pricing policy is an important component of the custodial evaluation. Global custody fees come in a wide variety of packages including: asset-based; transaction-based; account; income collection; delivery/receipt; reporting; administrative; miscellaneous (telephone, telex, facsimile, etc.); and other charges.

RFP respondents should be asked to clearly disclose and explain all charges. It may help to ask each custodian to provide a fee estimate based upon a hypothetical portfolio size, number of holdings, transactions, managers, and country distribution.

Account Management

Account management plays an important role in the global custody relationship, because it is the account manager that ties all of the component global custody services together for the client. Get to know the person who would have day-to-day responsibility for your account. It does no good to select the "best" custodian if the person handling your account is ineffective. A proactive, organized, and engaging enthusiasm for the account management role is essential. Look for low account management turnover and strong career pathing.

Conclusion

Today you are a prospective client and may demand all the trappings associated with the search and selection process. Tomorrow, however, you will be a client. The integrity and reputation of a bank's support for clients, as distinct from its supportiveness during the sales process, should be consciously and realistically evaluated. Seek exposure to people beyond the sales process. Remember that if a bank is offering you a "special deal," at the expense of its existing clients, this same tack may be taken when you are the client and they pursue other new business opportunities.

Avoid being distracted by inflammatory issues that draw attention away from core service components. Some custodians are struggling with the fundamentals. In an effort to mask those difficulties, they raise artificial issues or inflame relatively obscure points as major issues. If you are told of a "major issue," understand why it is important. What are the potential ramifications? Has there ever been a loss or problem? If so, why? If not, why not? Do not dwell on generalizations. Seek concrete examples and specifics.

The evaluation, selection, and monitoring of a global custodian is a difficult chore. Piercing the veneer of the sales process to understand the culture, commitment, and philosophy of business at a given organization can be critical in evaluating the organization and service it provides. Talk to people in the know. Consultants, investment managers, and clients all have had real-world exposure to a number of different global custodians and can provide verification and additional insight to what is learned in the formal search process. In sum, the careful and meticulous evaluation of all criteria in a balanced and considered way will help you choose the global custodian best able to meet your fund's particular needs.

Figure 1 **The Northern Trust Company: Purchase of Yen Securities and Foreign Exchange* on Initial or Subsequent Deposit of U.S. Dollars**

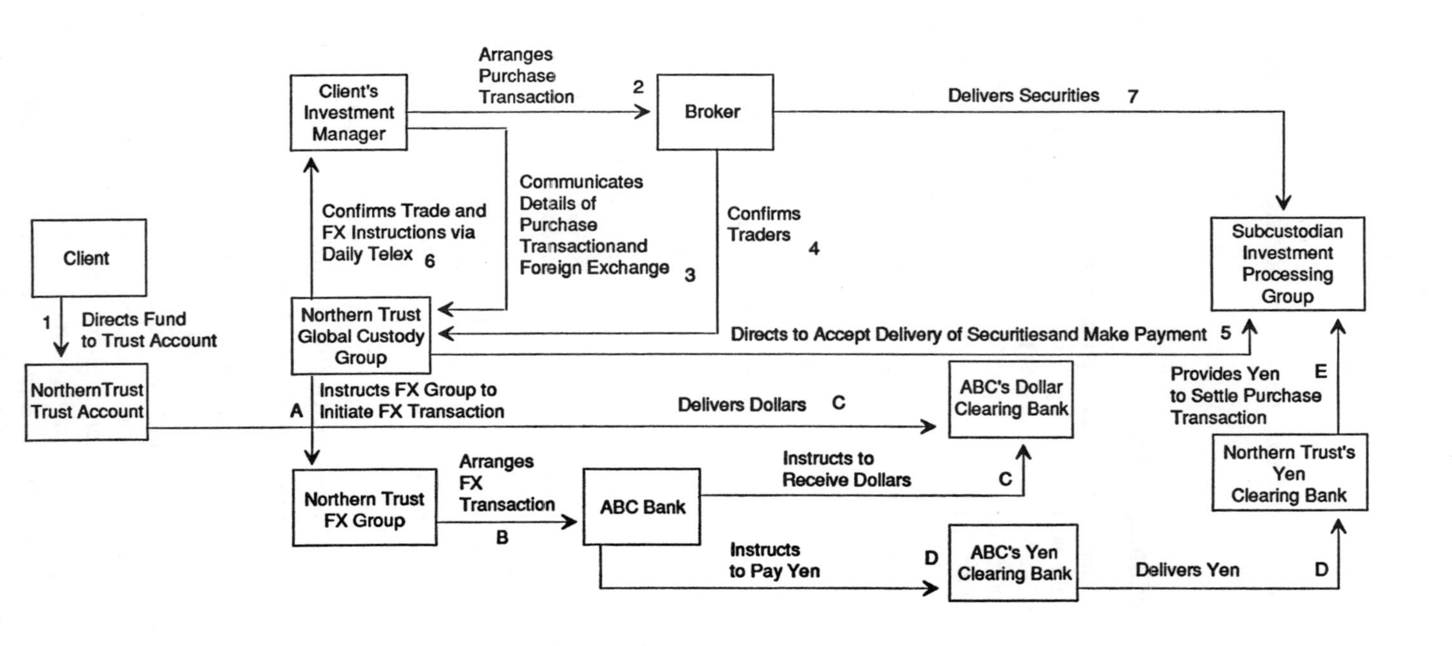

* With The Northern Trust Company

CHAPTER 15

Forecast Free International Asset Allocation*

H. Gifford Fong
Gifford Fong Associates

Introduction

Traditional asset allocation has typically taken the form of a mean/variance analysis.[1] This involves formulating a set of inputs, including expected returns, expected risk, a time horizon, and a set of investment-policy constraints. Each analysis includes for each asset class an expected return, standard deviation of return, correlation between each pair of asset classes, and perhaps a number of concentration constraints. These inputs are then analyzed by a quadratic optimization program, resulting in a series of efficient portfolios where there is a minimum risk combination of assets for each available return.

A portfolio that is diversified within a fixed-income portfolio may be thought of as the result of this same form of asset allocation, where the asset classes are typically short- and long-duration fixed-income assets. The same input considerations apply.

* This chapter is an adaptation of H. Gifford Fong and Oldrich A. Vasicek, "Forecast-Free International Asset Allocation," *Financial Analysts Journal*, March/April 1989. Copyright 1989 by the Financial Analysts Federation.

[1] Gifford Fong, "An Asset Allocation Framework," *Journal of Portfolio Management*, Winter 1980, pp. 58–66.

In a less formal approach, there may not be as much structure to the process. For example, many asset-allocation strategies are controlled by perceived historical patterns or relationships that generate expected returns. As with the more doctrinaire approach, the overriding reliance is on a set of return or interest-rate expectations. For those who have superior skills in accurately and reliably predicting rates or returns, these procedures will provide outstanding results.

There will be, however, groups of investors who do not believe they have superior forecasting ability, or there will be periods of time when expectations lack conviction. Moreover, the assumed normal return distributions of the expected returns may not conform to the desired outcomes of the investors. For these situations, an alternative path to asset allocation is appropriate.

What follows is an approach to asset allocation that is return and rate prediction free. It has the property that the return of the best-performing asset can be achieved, less some predetermined return differential, without having to predict which of the assets being used will do the best. Furthermore, the resulting return distribution can take on a variety of shapes, allowing flexibility in dealing with the investment objectives of the investor. This ability to deal with multiple-asset performance (MAP) is the essence of the strategy.

The next section describes the strategy in more detail. There follows a section on strategy implementation with a discussion of some simulation results. Finally, a brief summary concludes the chapter.

The Analysis

The objective of the analysis is to achieve a return on the total investment equal to the return of the best performing asset in a fixed-income portfolio, less the predetermined strategy cost. The strategy does not require any judgments about the expected returns of the individual assets to achieve this goal. In comparison with a mean/variance allocation, which provides a return equal to a weighted average of the separate asset returns, this strategy assures, after cost, the highest of the individual asset returns over the chosen investment horizon.

Consider, for example, an allocation of investment funds among U.S. stocks, U.S. bonds, and foreign equities. If U.S. stocks happen to perform the best of the three, the return on the strategy will be that of the U.S. stock portion less the known cost. If U.S. bonds perform better over the investment horizon than domestic and foreign stocks, the investment return will equal the return on bonds less some cost. If foreign stocks do better than domestic stocks or bonds, the investor will realize the return, after costs, of the foreign stock portfolio on his total investment.

The cost of the strategy increases with the number of assets involved. In addition to the number of assets, the cost depends on the riskiness of the assets (the riskier the assets, the higher the cost), the correlations among the assets (the more highly correlated they are, the lower the cost), and the length of the investment horizon (the cost per year decreases with increasing horizon length).

The strategy is implemented by a dynamic allocation of the investment funds among the several assets. The proportions of the total investment allocated to the individual assets are continuously monitored and adjusted, depending on their performance to date and on the time remaining to the horizon date. Of fundamental importance is the ability to achieve the best return without making any judgment on expected return.

Consider a strategy with an objective:

$$(1) \quad R_p = \max\,(R_1 - c_1, R_2 - c_2, ..., R_n - c_n),$$

where R_p is the total portfolio return, n is the number of assets in the asset allocation problem, R_1, R_2, ..., R_n are the individual asset returns, and c_1, c_2, ..., c_n are the strategy costs attributed to the individual assets. The goal is to get the best of multiple risky asset returns, less the corresponding cost. Thus, multiple assets such as domestic and foreign stocks, bonds, bills, and other asset types can be simultaneously considered. The strategy is equivalent to purchasing an option that allows the investor to choose the asset to call. Note that if one of the assets has a fixed return over the investment horizon, say R_n, then the strategy also guarantees a minimum return of $R_{\min} = R_n - C_n$.

For the two asset cases, the strategy cost can be determined from the Black/Scholes[2] option-pricing formula if interest rates are deterministic and constant, and from Merton's's[3] extension of that formula if the variability of interest rates is independent of their level. Margrabe[4] has provided a formula for two risky assets, and Stulz[5] derived an equation for two risky and one riskless asset. In general, the strategy cost will depend on the risk structure of the two assets throughout the horizon (i.e., their instantaneous covariance matrix as a function of time and state variables) and the length of the horizon.

The strategy costs c_1, c_2, ..., c_n in (1) are given by a valuation equation for the multiple-asset option. The valuation formula depends on the number of as-

[2] Fischer Black and Myron Scholes, "The Pricing of Options and Corporate Liabilities," *Journal of Portfolio Economy* 81, 1973, pp. 637–59.

[3] Robert Merton, "A Rational Theory of Option Pricing," *Bell Journal of Economics and Management Science* 4, 1973, pp. 141–83.

[4] W. Margrabe, "The Value of an Option to Exchange One Asset for Another," *Journal of Finance* 33, 1978, pp. 171–86.

[5] Rene Stulz, "Options on the Minimum or the Maximum of Two Risky Assets," *Journal of Financial Economics* 10, 1982, pp. 161–85.

sets, the covariance matrix of the n-dimensional stochastic process that characterizes each asset over the horizon, and the horizon length. For diffusion processes, the valuation formula involves $(n - 1)$-dimensional cumulative normal distribution functions with covariance matrices that are transformations of the n-dimensional instantaneous covariance matrix of the assets, integrated over the horizon.

Strategy Implementation

The strategy is executed by a dynamic allocation of investment funds among the several assets. The amounts allocated to the individual assets are maintained to be proportional to the partial derivatives with respect to the asset values of the valuation function (the same function that is also used initially to determine the costs of the strategy). The required allocation changes continuously as a function of the asset performance to date and the remaining time to the horizon.

As an example, consider n assets whose stochastic behavior is described by a logarithmic Wiener process. Let the instantaneous covariance matrix be specified by standard deviations all equal to 14 percent annual, with correlations among the assets all equal to .3, and assume a five-year horizon. The following table lists the value c of the uniform costs, in annual percent, as a function of the number of assets:

These costs are the price to pay for getting the best out of a number of asset returns. Suppose that the values of the parameters chosen for the example are descriptive of the international equity markets. It is possible to implement a strategy whose realized return is equal to the highest of six separate national stock markets, over a five-year period, less 5.7 percent annual. No prediction is needed as to which of these markets will have the highest return or, for that matter, what the expected returns are of each. In this type of strategy the investor attempts to capture the return variability of the underlying assets.

Let us look at a series of one-year, three-asset plans rolled over for three sequential years. Suppose in year one U.S. stocks returned 20 percent, Japanese stocks 5 percent, and U.K. stocks 5 percent, and the return differential is 8 percent, we would receive a return of 12 percent. If in year two, U.S. stocks returned 5 percent, Japanese stocks 20 percent, and U.K. stocks 5 percent, and the

Uniform Costs in Annual Percent

n	=	2	3	4	5	6
c	=	2.5	3.8	4.6	5.2	5.7

return differential is again 8 percent, we would obtain a return of 12 percent. If in year three, U.S. stocks returned 5 percent, Japanese stocks 5 percent, and U.K. stocks 20 percent, and the return differential is again 8 percent, we would again have a return of 12 percent. In this scenario the MAP portfolios would have an annual return of 12 percent, while each of the three assets had an annualized return of 10 percent. A static asset allocation over the three years would have the return of 10 percent, less than the 12 percent obtained by rolling over one-year MAP plans.

Table 1 and 2 present results of a study that was undertaken to determine the possible returns available had an investor applied the strategy utilizing three assets: the Standard & Poor's 500 Stock Index, Morgan Stanley's Japanese Stock Index (in U.S. dollars) and Morgan Stanley's United Kingdom Stock Index (in U.S. dollars), during the period January 1, 1978 to December 31, 1987.

Over the 10-year period, Japanese stocks performed extremely well. Japan turned in the best performance in six of the years and had an annual return of 26.5 percent. U.S. stocks performed best in three years and had an annual return of 15.3 percent. U.K. stocks were best in one year and had an annual return of 14.4 percent.

In the simulations, the risk parameters were estimated from the historical data for the three years prior to the inception date of the plans. Thus, the volatility estimates, return differentials, and initial hedge ratios are computed only with the information available at the time each plan was started.

The simulation results for the 10 one-year plans are shown in Table 1. Over the 10-year horizon Japan's returns varied between –13.7 percent and 98.2 percent with an annual return of 26.5 percent and a standard deviation of 29.7 percent. The MAP portfolio returns ranged from 2.0 percent to 75.8 percent with an annual return of 19.4 percent and a standard deviation of 20.2 percent. The MAP portfolios captured a majority of the best performer's returns, had a much lower standard deviation and never lost money.

When comparing the simulation results to the Morgan Stanley World Equity Index, the results are even more encouraging. On an annual basis, the MAP portfolios were able to add 500 basis points per year over the World Index. Had an investor invested $1 million in the World Index ten years ago, it would have grown to $3.8 million. The same $1 million invested in the MAP portfolio would have grown to $5.9 million. All this while never incurring a principal loss.

The simulation results for the five two-year plans are shown in Table 2. The series of five MAP portfolios had an annual return of 19.0 percent and an annual return of 14.3 percent and an annual standard deviation of 13.9 percent.

Table 1 International Equity Portfolio Annual Returns, One-Year Plans (1978–87)

	U.S.[a] (%)	Japan[b] (%)	U.K.[c] (%)	MAP[d] (%)	World[e] (%)
1978	6.47	50.08*	8.56	29.60	12.68
1979	18.47*	–13.65	15.56	9.14	7.21
1980	32.40*	27.74	32.36	19.58	21.46
1981	–5.01	13.84*	–15.93	2.02	–7.92
1982	21.56*	–2.28	3.11	6.54	5.82
1983	22.59	23.05*	11.77	14.86	18.56
1984	6.39	15.73*	0.64	6.14	3.25
1985	31.80	41.87	46.31*	25.58	37.02
1986	18.71	98.15*	21.94	75.81	39.10
1987	5.21	42.56*	32.70	18.26	14.30
Geo. Mean	15.27	26.52	14.38	19.35	14.32
Std. Dev.	11.61	29.68	17.31	20.19	13.91

[a] Standard & Poor's 500.
[b] MSCI's Japanese Stock Index (in U.S. dollars).
[c] MSCI's U.K. Stock Index (in U.S. dollars).
[d] Multiple asset performance strategy.
[e] MSCI's World Stock Index (capitalization-weighted).
* Denotes highest return for each period.

Table 2 International Equity Portfolio Annual Returns, Two-Year Plans (1978–87)

	U.S.[a] (%)	Japan[b] (%)	U.K.[c] (%)	MAP[d] (%)	World[e] (%)
1978–79	12.31	13.84*	12.00	6.13	9.91
1980–81	12.31	20.57*	5.48	10.93	5.75
1982–83	22.08*	9.65	7.35	12.50	12.01
1984–85	18.39	28.09*	21.31	16.80	18.94
1986–87	11.76	68.07*	27.21	54.38	26.09
Geo. Mean	15.27	26.52	14.38	19.01	14.32
Std. Dev.	11.61	29.68	17.31	21.51	13.91

[a] Standard & Poor's 500.
[b] MSCI's Japanese Stock Index (in U.S. dollars).
[c] MSCI's U.K. Stock Index (in U.S. dollars).
[d] Multiple asset performance strategy.
[e] MSCI's World Stock Index (capitalization-weighted).
* Denotes highest return for each period.

Conclusion

The study demonstrates that one can dynamically capture the variability of a group of asset returns. More important, this strategy can add significant value over a benchmark if there is high variability in the return among the different asset classes. While many active managers attempt to forecast expected returns or rates, this is a nonexpectational model that uses options-pricing theory to achieve the return of the best performing asset less a return differential. By capturing the assets' return variability, these portfolios may actually outperform all the underlying assets after a few years.

Index

A

B

C

D

E

F

G

J-K

L

M

N

O

P-Q

R

S

T

U-V

W

Y